Springer Texts in Business and Economics

Springer Texts in Business and Economics (STBE) delivers high-quality instructional content for undergraduates and graduates in all areas of Business/Management Science and Economics. The series is comprised of self-contained books with a broad and comprehensive coverage that are suitable for class as well as for individual self-study. All texts are authored by established experts in their fields and offer a solid methodological background, often accompanied by problems and exercises.

More information about this series at http://www.springer.com/series/10099

Thorsten Hens • Marc Oliver Rieger

Solutions to *Financial Economics*

Exercises on Classical and Behavioral Finance

Springer

Thorsten Hens
Department of Banking and Finance
University of Zurich
Zürich, Switzerland

Marc Oliver Rieger
Department of Business Administration
University of Trier
Trier, Germany

ISSN 2192-4333 ISSN 2192-4341 (electronic)
Springer Texts in Business and Economics
ISBN 978-3-662-59887-0 ISBN 978-3-662-59889-4 (eBook)
https://doi.org/10.1007/978-3-662-59889-4

This Springer imprint is published by the registered company Springer-Verlag GmbH, DE, part of Springer Nature.
The registered company address is: Heidelberger Platz 3, 14197 Berlin, Germany

Preface

Writing a textbook takes time and effort. Everybody knows that. But designing exercises is by no means an easier task as we have noticed when composing this book. Good exercises are challenging and enlightening and even fun to solve! It's not easy to find them. We hope that we were able to find such exercises for this book, if not always, at least sometimes. A few times, however, others have found so wonderful exercises for a topic that it is difficult to find something better and we have therefore—with permission of the authors—reprinted classical exercises from other books.

The structure of the book is self-explanatory: in the first part you find exercises and in the second part solutions. The chapter numbers follow the chapters of our textbook *Financial Economics* to make it easy to navigate. We hope you will learn a lot from these exercises, we also hope that you like them and we appreciate the effort you take studying them!

We thank Marie Hardelauf for the typing and the layout of the book, Artem Dyachenko for his input to some of the exercises, and Anastasiia Sokko, Nilüfer Schindler, Urs Schweri and Sabine Elmiger for their general help with composing this book. We thank Biljana Meiske for recomputing and improving many exercises. Finally, we thank Philipp Baun and his team from the Springer publishing house for their immense patience with us and their steady support.

Zürich, Switzerland
Trier, Germany
July 2019

Thorsten Hens
Marc Oliver Rieger

Contents

Part I

Exercises

1 Introduction

In this part, we present a large number of exercises that can accompany our book *Financial Economics*. They are sorted by the chapters of the book. Within each chapter, the exercises are roughly sorted by topics such that topics covered earlier in the chapter come first. A second criterion is by the difficulty (the easier exercises first). Many exercises are far from being "routine". We think that exercises that just plug in numbers into formulas being learned by heart do not help students much to comprehend a topic. They also don't help the lecturer when designing exams or homework assignments: such simple "plug-and-play" exercises are easier designed from scratch then copied from a textbook. Instead, we tried to design exercises that inspire thinking and encourage deeper understanding of the subject. Often there is not only one solution and sometimes students who do not find the optimal solution can at least try to get partial or approximative results. Teasing out the creativity of students in solving problems is very helpful for guiding them into making their own research and this is what we aim to achieve with our exercises.

That does not mean that our exercises are all very difficult and only solvable for top students. To the contrary, we hope that most students who put enough thought and effort into them will be able to solve them—at least partially.

Finally, we hope that the exercises do not only encourage thinking and help to understand the topics of our book, but also are interesting—and sometimes even fun—to solve!

Complete solutions to all exercises are given in the second part of this book.

T. Hens, M. O. Rieger, *Solutions to* Financial Economics,
Springer Texts in Business and Economics,
https://doi.org/10.1007/978-3-662-59889-4_1

2 Decision Theory

2.1. Consider the following game: you roll a dice, if you roll a 6, you win 6 million € otherwise you win nothing. You can play only once. Let us assume your expected utility function is given by $u(x) = \log_{10} x$ (base 10 logarithm, i.e., $\log_{10}(10^n) = n$) and your initial wealth is 10,000 €.

(a) How big is your expected utility after playing this game? Imagine instead that you get 1 million € for sure, how big is your utility afterwards? Which of the two variants would you therefore prefer? How could you have seen this without doing any computation?
(b) Now, the prize of the game is only 61 €. What would be the certainty equivalent of the game, given the same expected utility function as above? Should you participate for a fee of 10 €? Why is the result surprising?

2.2. Prove that the Expected Utility Theory (EUT) satisfies the Continuity Axiom.

2.3. In a city center, parking space is rare. Hence, legal parking costs an amount of $t > 0$. Some people decide to park illegally. There is a probability $p > 0$ of being caught which leads to a fine $f > t$. In order to decrease the number of illegal parkers, there are two possible concepts: doubling the fine f or doubling the controls (i.e., the probability p). Assuming that the illegal parkers are risk-averse, which is the better concept?

2.4. Consider two assets: a stock and a bond. There are two states of the world (each with probability 1/2): boom and recession. The stock's returns are +8% in a boom and −2% in a recession, the bond yields +2% each.

(a) Compute their mean and variance.
(b) Find the value of α such that an investor with the mean-variance utility function $U(\mu, \sigma) = \mu - \alpha\sigma^2$ is indifferent between both assets.

T. Hens, M. O. Rieger, *Solutions to* Financial Economics,
Springer Texts in Business and Economics,
https://doi.org/10.1007/978-3-662-59889-4_2

(c) If this investor buys some stocks (say a proportion $\lambda \in [0, 1]$ of his total investment) and some bonds (a proportion of $1 - \lambda$), how will his returns be distributed? Which $\lambda \in [0, 1]$ is optimal for him?

2.5. Daniel Bernoulli, one of the founders of expected utility theory, and Daniel Kahneman, one of the founders of Prospect Theory, go on vacation. They each have two credit cards and two wallets. With a certain probability a wallet could be stolen. The probability that a particular wallet is stolen is independent from the probability that another wallet is stolen.

Assume that both act according to their theories. Would they put both credit cards into the same wallet or each in a different wallet? [Hint: Assume an identity probability weighing function, i.e., $w(p) = p$.]

2.6. Can the standard form of PT with the standard PT-parameters explain that people play a lottery if the winning probability is 1:1,000,000, the prize is one million € and a lottery ticket costs 2 €?

2.7. Show that Cumulative Prospect Theory explains Allais' Paradox. To this aim, compute the CPT values of the four lotteries and compare!

Lottery A	State	1–33	34–99	100
	Outcome	2500	2400	0
Lottery B	State	1–100		
	Outcome	2400		
Lottery C	State	1–33	34–100	
	Outcome	2500	0	
Lottery D	State	1–33	34–99	100
	Outcome	2400	0	2400

2.8. Can the certainty equivalent of a lottery in Prospect Theory (PT) be larger than the largest outcome of the lottery? How is it in Cumulative Prospect Theory (CPT)? How is it in the normalized version of Prospect Theory (by Karmarkar)? Give an example or proof! (This property is called "violation of internality".)

2.9. We say that a person is loss averse if he does not like to participate in a lottery with 50% chance of winning X and 50% chance of losing X. Let us assume a person's decisions are described by classical prospect theory with parameters $\alpha < \beta$. For simplicity, assume $X = 100$, $\alpha = 0.8$, $\beta = 1$, i.e. risk neutrality in losses, and $\gamma = 1$, i.e. no probability weighting.

Compute the values of λ for which the person is loss averse! Show that for any $\alpha < \beta < 1$ the person can be loss averse for some $\lambda < 1$!

2.10. Let us assume that a value function v is given by $v(x) := x$ and a weighting function w is $w(F) := \sqrt{F}$. A lottery is described by the probability measure

$P := p(x)\,\mathrm{d}x$, where the probability density p is given as

$$p(x) := \begin{cases} x, & \text{if } 0 \leq x < 1, \\ 2 - x, & \text{if } 1 \leq x < 2, \\ 0, & \text{otherwise.} \end{cases}$$

Compute the CPT-value of this lottery. Use this to compute the certainty equivalent (CE). Explain the difference between the CE and the expected value!

2.11. Jerome is a student. If you ask him whether he prefers 100 € now or 110 € next week, he prefers to get the money now. If you ask him, however, whether he prefers 100 € now or 200 € in 4 months, he prefers to wait. Can you explain these preferences with classical time discounting? Can you explain it with hyperbolic discounting? Assume linear utility e.g. $U(w) = w$.

Angelika is a student. If you ask her whether she prefers 100 € now or 120 € next year, she prefers to wait. If you ask her, however, whether she prefers 100 € now or 1000 € in 10 years she prefers the money now. Can you explain these preferences with classical time discounting? Can you explain it with hyperbolic discounting? What if you consider an increase in her wealth level in 10 years? [Hint: assume a non-linear utility function e.g. $U(w) = \min(w, c)$ where c is a constant.]

2.12 (Samuelson Paradox). We all know that we can take more risk in our investment decisions when we have a longer investment horizon—do we? Consider the following counter argument by Paul Samuelson: let us suppose you are not willing to play a certain gamble only once, but you are willing to accept the offer to play it ten times. Now, after playing it nine times, why don't you want to stop here? After all, past is past, and at this point you just have to decide to play this gamble *once* (more) or not and you preferred in this case not to play it, didn't you? So, you would rather only play nine times. But then of course the same argument could be iterated and you would finally not play the gamble at all. Now replace "gamble" by "investing in the stock market for one year" and you have just disproved that you should be willing to take more risk on the long run.

On the other hand, if you choose a utility function, say, $u(x) = x^{\alpha}$, you can construct a lottery L such that the utility of this lottery is lower than the utility of not playing, but the utility of playing the lottery twice (or ten times) is larger than not playing. (Construct such a lottery as an exercise!) So this tells you that, yes, indeed a rational person might want to take more risk on the long run.

Now we have two nice proofs that contradict each other, a situation we tend to call a paradox.

> How wonderful that we have met with a paradox. Now we have some hope of making progress

we could say in the words of Niels Bohr. But how do you solve this paradox?

2.13. There are three assets. Their payoffs are as follows:

Probability	Asset 1	Asset 2	Asset 3
50%	5%	5%	– 5%
50%	10%	–5%	–10%

Assume $R_f = 0\%$.

(a) Calculate the certainty equivalent of each asset, if the investor has the utility function

$$v_{KT}(R) = \begin{cases} (R - RP)^{\alpha^+}, & \text{if } R > RP, \\ -\beta\,(RP - R)^{\alpha^-}, & \text{if } R \le RP, \end{cases}$$

with $\alpha^+ = \alpha^- = 0.88$, $\beta = 2.25$ and $RP = 0\%$. The values for the alphas and beta are the classical results of Kahneman and Twersky.

(b) Find the parameters of

$$v_{DHM}(R) = \begin{cases} (R - RP) - \frac{\alpha^+}{2}(R - RP)^2, & \text{if } R > RP, \\ \beta\left((R - RP) - \frac{\alpha^-}{2}(R - RP)^2\right), & \text{if } R \le RP, \end{cases}$$

such that the three assets have the same certainty equivalent as in the v_{KT}-case (assume $RP = 0\%$ also in v_{DHM}).

2.14. In a market we have two states, a risky and a risk-free asset. State 1 occurs with a probability $p = 75\%$. The risky asset pays 7% in state 1 and –10% in state 2. The risk-free asset pays 1% for sure.

(a) Determine the optimal portfolio of a mean-variance maximizer with $\gamma = 2$.

(b) Determine the optimal portfolio of a prospect theory maximizer with

$$v(R) = \begin{cases} (R - RP) - \frac{\alpha^+}{2}(R - RP)^2, & \text{if } R > RP, \\ \beta\left((R - RP) - \frac{\alpha^-}{2}(R - RP)^2\right), & \text{if } R \le RP, \end{cases}$$

with $\alpha^+ = -\alpha^- = 1.5$, $\beta = 2$ and $RP = R_f$. Assume that the agent is not doing probability weighting. Assume there are short selling constraints i.e. $\lambda \in [0, 1]$.

(c) Assume the stock has now a return of –2% in state 1 and 16% in state 2. Determine the optimal portfolio of the same investor but he weights the probabilities as follows: $w(0.25) = 0.3$ and $w(0.75) = 0.7$.

2.15.

(a) Suppose you have a 50% chance to double your yearly income. Which percentage of your yearly income are you willing to risk in the other 50% of the cases?

(b) Assume you have an expected utility function with constant relative risk aversion. Compute the size of your CRRA.

(c) Compute your risk aversion in the mean-variance approach. The utility function is $U(R) = \mu(R) - \gamma/2\,\sigma(R)^2$.

(d) Which percentage of income would the following representative prospect theory agent of Kahneman and Tversky with

$$v(R) = \begin{cases} (R - RP)^{\alpha^+}, & \text{if } R > RP, \\ -\beta\,(RP - R)^{\alpha^-}, & \text{if } R \leq RP, \end{cases}$$

$\alpha^+ = \alpha^- = 0.88$, $\beta = 2.25$, $RP = 0\%$ and without probability weighting risk in that situation?

(e) Find the risk aversion of a mean-variance investor such that he would split his wealth equally between stocks and bonds. To do so recall that the equity premium is 6.4% and the standard deviation of stocks is 21%.

(f) Now suppose you have some background wealth which is 50% of your yearly income. Take the percentage of answer (a) and find your CRRA for this case.

3 Two-Period Model: Mean-Variance Approach

3.1. There are two risky assets, $k = 1, 2$ and one risk-free asset with return of 2%. Risky assets cannot be short sold. The expected returns of the risky assets are $\mu_1 := 5\%$ and $\mu_2 := 7.5\%$. The covariance matrix is:

$$COV := \begin{pmatrix} 2\% & -1\% \\ -1\% & 4\% \end{pmatrix}.$$

(a) Calculate the Minimum-Variance Portfolio and the Tangent Portfolio.

(b) Some mean-variance investor assuming the Covariance Matrix given above chooses the portfolio $\lambda := (0.2, 0.5, 0.3)$. Assume that investor's risk aversion is $\rho := 1$. Which implicit expected returns does he hold?

(c) Suppose the market portfolio is $\lambda^M := (0.4, 0.6)$. Compute the beta-factors. Assume the excess return of the market portfolio is 3%. Determine the expected returns of the two risky assets.

3.2. Download the data from the homepage of this book (you find it at http://www.financial-economics.de). The datafile contains the price levels of equity indices of several countries as well as the world risk-free interest rate, which represents the monthly returns of the risk-free asset. Prepare the country index data by calculating the monthly net returns first.

(a) Compute the mean and standard-deviation of the monthly net-returns of the indices given. Determine the corresponding annual returns. Is there differences in the risk-return tradeoff across different country equity markets? Interpret briefly.

(b) Plot the histograms of the monthly returns. Explain the distributional properties of the country returns such as possible fat tails or skewness.

T. Hens, M. O. Rieger, *Solutions to* Financial Economics,
Springer Texts in Business and Economics,
https://doi.org/10.1007/978-3-662-59889-4_3

(c) Compute the covariance/correlation matrix. What country indices are highly correlated? Are there any negative correlations?
(d) Draw the efficient frontier for the asset universe given in the data set. Show both the unconstrained and the constrained case, where no short sales are allowed.
(e) Compute the tangent-portfolio with and without short-sales. Are they different? Explain why.
(f) Illustrate the sensitivity of the tangent portfolio with respect to changes in means by drawing different frontiers (unconstrained).
(g) Compute the SML based on the tangent-portfolio with short-sales. Show that all the assets are on the SML graphically.

3.3. Construct a simple example with three risky assets $k = 1, 2, 3$ such that none of them have a pairwise correlation of $+1$ or -1, but a combination of asset 1 and 2 has a correlation of $+1$ with asset 3.

Use this example to prove that the problem described in Sect. 3.1.4 of the book cannot be solved by controlling for pairwise correlation between assets.

3.4. An investor with mean-variance utility $U(\mu, \sigma) := \mu - \sigma^2$ can invest in three risky assets, $k = 1, 2, 3$ and one risk-free asset. The risk-free return is 2%. Risky assets cannot be short sold. The expected returns of the risky assets are $\mu_1 := 5\%$, $\mu_2 := 7.5\%$ and $\mu_3 := 10\%$ The covariance matrix is:

$$COV := \begin{pmatrix} 2\% & -1\% & -2\% \\ -1\% & 4\% & 6\% \\ -2\% & 6\% & 8\% \end{pmatrix}.$$

(a) Calculate the optimal portfolio λ^{opt} of the investor if he can only invest in the first two assets. Calculate the mean and the variance and also the investor's utility for that portfolio.
(b) Now consider the third asset and show that it has positive Alpha with respect to the optimal portfolio λ^{opt}. Suggest a new portfolio mix consisting of the tangential portfolio and the third asset so that the investor improves upon the tangential portfolio.
(c) Now suppose the investor had initially chosen the portfolio consisting of asset 2 only. Show that adding asset three to this portfolio worsens his situation!

3.5. Suppose that financial markets consist of two risky assets and one riskless asset. Let $R_f = 1\%$ and there be four investors each of whom has different beliefs for the expected returns of the two risky assets as follows:

$$\mu^1 = \begin{pmatrix} 6\% \\ 1\% \end{pmatrix}, \quad \mu^2 = \begin{pmatrix} 3\% \\ 2\% \end{pmatrix}, \quad \mu^3 = \begin{pmatrix} 2\% \\ 3\% \end{pmatrix}, \text{ and } \mu^4 = \begin{pmatrix} 1\% \\ 5\% \end{pmatrix}.$$

All investors have the same degree of risk aversion in the mean-variance preferences, $\rho^1 = \rho^2 = \rho^3 = \rho^4 = 2$, and their invested wealth levels are all the same as well $w_0^1 = w_0^2 = w_0^3 = w_0^4 = 10$. The variance-covariance matrix and the true expected returns of the risky assets are given by

$$COV = \begin{pmatrix} 2\% & 0\% \\ 0\% & 2\% \end{pmatrix} \text{ and } \hat{\mu} = \begin{pmatrix} 2\% \\ 2\% \end{pmatrix}.$$

(a) Find which of the investors should be actively investing and which should rather not.
(b) Show that investor 2 has a negative alpha portfolio but should rather be active.
(c) Show that investor 4 has a positive alpha portfolio but should rather be passive.

3.6. "Equities, bonds and other traditional asset classes have an economic rationale for giving positive mean returns. Hedge funds have no economic theory underlying their positive performance. There is no risk premium in the classic economic sense. The returns are achieved by the managers' ability to exploit inefficiencies left behind by other (less informed, less intelligent, less savvy, ignorant, or uneconomically motivated) investors in what is largely considered a zero or negative sum game." *Alexander M. Ineichen (UBS Investment Research, March 2005, page 31.)*
In the following we analyze this statement critically:
Consider a two-period financial market model with $k = 0, 1, \ldots, K$ assets and heterogeneous beliefs. Let $k = 0$ be the risk-free asset.

(a) For the CAPM, define the ex-post alpha of an asset k, $\widehat{\alpha}_{k,M}$, and of an investment strategy $\lambda^i = (\lambda^{i1}, \ldots, \lambda^{iK})$, denoted by $\widehat{\alpha}^i$.
(b) Let r^i be the relative wealth of investment strategy λ^i. Argue that $\sum_i \widehat{\alpha}^i r^i = 0$, i.e., with respect to the alphas financial markets are a zero sum game.
(c) In the last 10 years Hedge Funds have generated positive returns of about 10% p.a. Is this finding compatible with the CAPM?
(d) Comment on the quotation from Ineichen given above. Are his statements supported by financial economics as it has been taught in this class?

3.7. Let $R_f := 1\%$ and let there be two risky assets and three investors with the following characteristics:

$$\mu^1 := \begin{pmatrix} 3\% \\ 1\% \end{pmatrix}, \quad \mu^2 := \begin{pmatrix} 2\% \\ 1\% \end{pmatrix}, \quad \mu^3 := \begin{pmatrix} 1\% \\ 2\% \end{pmatrix},$$

$$\gamma^1 := \gamma^2 := \gamma^3 := 2,$$

$$w_0^1 := w_0^2 := w_0^3 := 5.$$

Let

$$\operatorname{cov} := \begin{pmatrix} 2\% & 0\% \\ 0\% & 2\% \end{pmatrix} \text{ and } \hat{\mu} = \begin{pmatrix} 2\% \\ 1\% \end{pmatrix}.$$

(a) Calculate the (ex-ante) market expectation $\overline{\mu}$.
(b) Calculate the optimal portfolio for all investors (if they are active).
(c) Calculate the market portfolio λ_M assuming that all investors are active.
(d) Which investors should invest active, which passive?
(e) Calculate the ex-post alphas of the investors.
(f) Show that investor 1 has a positive ex-post alpha, if he is active, but should better be passive.

3.8. Consider two risky assets with

$$(\mu_1, \sigma_1) := (5\%, 5\%) \qquad \text{and}$$
$$(\mu_2, \sigma_2) := (10\%, 10\%).$$

The correlation between the two assets is $\rho = 0.5$.

(a) Compute the tangent portfolio for $R_f = 0\%$ with and without short-selling.
(b) How does the tangent portfolio change when R_f increases?

4 Two-Period Model: State-Preference Approach

4.1. Consider a two-period economy with uncertainty in the second period. Consumption is in terms of a single consumer good. In the second period there are **S** many possible states and every consumer aims to maximize the consumption across states. There are **I** many consumers with utility functions $\mathbf{U}^i$(strictly increasing, concave and continuous). The consumption good has a price π_s in each state. Consider the following market structures:

(A) Arrow Debreu Equilibrium: pure exchange, consumption c_s^i, endowment ω_s^i with prices π_s.

(B) Financial Markets Equilibrium: consumption c_s^i, endowment ω_s^i with an economy where exchange is done solely via financial markets. There are $\mathbf{K} + 1$ many financial assets in the economy with period 1 prices q^k and each asset has **S** many possible payoff scenarios A_s^k in the next period and agents transfer their consumption via asset allocations $\theta^{i,k}$.

(a) State the budget constraints of the consumers in both economies A and B.

(b) State the market equilibrium for both economies A and B.

(c) Assume that the financial markets are complete and show that the equilibrium of the economies A and B will yield the same consumption allocations.

4.2. Suppose we have financial markets with the following payoff matrix and market equilibrium prices

$$A = \begin{pmatrix} 1 & 4 \\ 1 & 2 \\ 1 & 0 \end{pmatrix} \quad q = (1, 2)$$

T. Hens, M. O. Rieger, *Solutions to* Financial Economics,
Springer Texts in Business and Economics,
https://doi.org/10.1007/978-3-662-59889-4_4

Introduce a new asset with a payoff (2 0 0). This can be a call option written on the second asset with a strike price of 2. Find the price or pricing bounds using the no-arbitrage condition.

4.3. Suppose we have a financial market with the following payoff matrix of two assets in two states:

$$A = \begin{pmatrix} 1 & 3 \\ 1 & 1 \end{pmatrix}$$

(a) Find which of the following set of prices are arbitrage-free or not by trying to form arbitrage strategies:
  (i) $q' = [1,\ 2]$
  (ii) $q' = [1,\ 1/2]$
  (iii) $q' = [1,\ 1]$

(b) Find which of the following set of prices are arbitrage free or not by calculating the state prices implied by *FTAP*, conclude which set of state prices are positive for which set of asset prices.
  (i) $q' = [1,\ 2]$
  (ii) $q' = [1,\ 1/2]$
  (iii) $q' = [1,\ 1]$

(c) Consider the asset prices that are arbitrage-free and introduce a call option which has a strike price of 1.5 on the second asset in the economy.
  (i) State the payoff of the call option.
  (ii) Is it possible to replicate the payoff of the call option?
  (iii) Calculate the price of the call option by considering the arbitrage free prices given in the previous parts.

4.4. Consider an economy with two agents, two equally likely states $\Rightarrow S = 2$, each agent has the expected utility of the form:

$$U^1(c) = U^2(c) = \ln(c_0) + \frac{1}{2}(\ln(c_1) + \ln(c_2))$$

and endowments

$$w^1 = (1, 1, 2),$$
$$w^2 = (1, 2, 1).$$

In financial markets, suppose we have a single bond paying one in both states.
Then the payoff matrix is $A' = [1\ 1]$

(a) Is the market of this economy complete?

(b) Derive the financial markets equilibrium asset and consumption allocations and asset prices. Show whether there is trade among consumers.

(c) Is there a way to complete this market? What kind of payoff vector of an asset would complete the market? Give an example.

(d) Consider a new asset with the payoff vector $[x,\ y] \in \mathbb{R}^2$ where $x \neq y$, where $x = 2$ and $y = 1/2$. Can you find the price of this asset? Otherwise, find the valuation bounds for this asset.

(e) Introduce a new asset to the market with the payoff vector $[x,\ y] \in \mathbb{R}^2$ where $x \neq y$. Now that the market is complete, derive the equilibrium allocations and prices.

(f) Find the first asset's equilibrium price in the new financial market economy and compare whether the introduction of the new asset changes the existing asset's equilibrium price.

(g) Suppose the payoff vector of the second asset as it given in part (d). $[x,\ y] \in \mathbb{R}^2$ where $x \neq y$, where $x = 2$ and $y = 1/2$. Find the equilibrium price of the asset in the new financial markets equilibrium. Compare if the new asset belong to the valuation bounds you found in part (d).

(h) With the introduction of the second asset, did the asset allocations of the consumers change?

4.5. Suppose we have financial markets with a payoff matrix

$$\begin{pmatrix} \text{Bond} & \text{Stock} \\ 1 & 3 \\ 1 & 1 \\ 1 & 1 \end{pmatrix}$$

(a) Introduce a call option to this market with $K = 1$

$$\text{Payoff} = \max(0, S - K) \Rightarrow \begin{bmatrix} 2 \\ 0 \\ 0 \end{bmatrix}$$

What happens to the market? Can we price the call option with bond and stock? Is it a redundant asset?

(b) Assume now with the new asset that we have three assets in the markets as in part (a). Introduce a second call option with a different strike price $K = 2$.

$$\text{Payoff} = \max(0, S - K) \Rightarrow \begin{bmatrix} 1 \\ 0 \\ 0 \end{bmatrix}$$

Show whether the second option becomes redundant and if we can replicate the option payoffs.

(c) Introduce a put option (as a second option) $K = 1$ in the markets with 3 assets, 1 bond, 1 stock and 1 call option.

$$\text{Payoff} = \max(0, K - S) \Rightarrow \begin{bmatrix} 0 \\ 0 \\ 1 \end{bmatrix}$$

How can we price this asset? Derive the put-call parity by using the replicating strategy method.

4.6. Consider an economy with two consumers, two time periods, two equally likely future states and financial markets with 2 assets. Each agent has the following expected utility function form

$$U^i(\mathbf{c}) = \ln(c_0^i) + \frac{1}{2}\ln(c_1^i) + \frac{1}{2}\ln(c_2^i)$$

and initial endowments $\omega^1 = (1,\ 1,\ 2)$ and $\omega^2 = (1,\ 2,\ 1)$. The financial markets have two assets in zero net supply with the following payoff matrix:

$$\begin{pmatrix} 1 & 1/2 \\ 1 & 2 \end{pmatrix}$$

(a) Check whether the markets are complete.
(b) Define the financial market equilibrium for this economy.
(c) Define the Arrow Debreu equilibrium for this economy.
(d) Find the equilibrium state prices.
(e) Find the financial assets' prices by using the state prices.
(f) Find the optimal consumption allocations for both agents.
(g) To derive the optimal asset allocations, use the budget constraints of the financial markets equilibrium as both equilibrium solution would yield the same consumption allocation. We could use the asset allocations that would achieve the desired optimal consumption allocation we derived in the Arrow Debreu equilibrium.
(h) Is there any wealth transfer across time? Explain why.
(i) Is there wealth transfer across states? Explain why for each agent in detail.
(j) In the economy, we introduce a new asset called a derivative. It is a call option written on the risky asset in the economy with the second period exercise price $K = 1.1$. What would be the price of this option...
 (i) by using the hedge portfolio method, by deriving the amount to invest in both assets riskless and risky?
 (ii) by using the equilibrium state prices?
 (iii) by using the risk neutral pricing formula?

4.7. A consumer has to decide between consumption now, c_0, and consumption tomorrow, c_1. The income of the consumer today is $w_0 = 5$ and the income in the next period is $w_1 = 20$. The price of the consumption is in both periods 1. The consumer may save the income of today (at zero interest) for the next period, but he cannot borrow money. The utility function of the consumer is $u(c_0, c_1) = c_0c_1 + c_1$.

(a) Formulate the maximization problem.
(b) What is the optimal consumption plan of the consumer?
(c) Now the consumer can borrow (or lend) money at an interest rate $r = 5\%$. What is his optimal consumption plan in this case?
(d) At which interest rate r_0 is the consumption plan identical to the one in (b)?

4.8. An investor has an investment budget of $w = 100$, which he can spend today. He can invest into stocks from Shell and GM. There are two states in the world, in state 1 the oil price is high and in state 2 the oil price is low. The payment of the stocks depending on the states is:

	Shell	GM
Oil high	5	1
Oil low	2	4
Price	1	1

The utility function of the investor is $u(c_{high}, c_{low}) = \log(c_{high}) + \log(c_{low})$. How many stocks of each kind will the investor buy?

4.9 (Intertemporal Trade—Wealth Transfer Across Time). Consider a two-period economy with a representative agent and without uncertainty, i.e., we have $S = 1$ for the number of states in the second period. In both periods the agent receives income $w = (w_0, w_1)$, which is growing at the rate g. i.e., $w_1 = gw_0$.

The representative agent is a utility maximiser with

$$U(c) = u(c_0) + \delta u(c_1),$$

where δ is the discounting factor and and c_0, c_1 his consumption in the first and second periods correspondingly. He has a possibility to borrow and save money at the interest rate r.

(a) Formulate the decision problem of the representative agent.
(b) Assume that the utility function is of type $u(c) = \ln(c)$. What is the optimal saving/borrowing amount s? Under which conditions would the representative agent save, i.e., $s > 0$? Explain the implications of the discounting factor, income growth and interest rate on the optimal saving/borrowing amount s.

4.10. Consider a financial market with two assets and two states. Their payoffs are given by

$$A = \begin{pmatrix} 100 & 50 \\ 100 & 200 \end{pmatrix},$$

and their prices by $q = (100, 100)'$. What is the price of a call option on asset 2 with strike K? What is the price of a put option on asset 2 with strike K?

4.11. Consider a market with two assets and three states, each with a probability of $1/3$. Let the payoffs be given by the matrix

$$\boldsymbol{A} := \begin{pmatrix} 1 & 1 \\ 1 & 1.5 \\ 1 & 0.5 \end{pmatrix}.$$

(a) Explain why a security with payoffs $(1, 0, 0)'$ cannot be replicated by the assets.
(b) Assume the asset prices are given by $q = (1, 1)'$. Find the bounds for its price from below and above.

4.12. Consider four financial markets with the following payoff matrices and price vectors:

$$A_1 = \begin{pmatrix} 2 & 1 \\ 0 & 1 \end{pmatrix}, \quad A_2 = \begin{pmatrix} 1 & 4 \\ 1 & 2 \\ 1 & 0 \end{pmatrix}, \quad A_3 = \begin{pmatrix} 2 & 2 & 2 \\ 2 & 0 & 1 \\ 0 & 2 & 1 \end{pmatrix}, \quad A_4 = \begin{pmatrix} 2 & 1 & 4 \\ 2 & 2 & 1 \\ 2 & 4 & 1 \end{pmatrix},$$

$$q^1 = (q_1^1, q_2^1), \quad q^2 = (1, 2), \quad q^3 = (q_1^3, q_2^3), \quad q^4 = (1, 1, 1).$$

(a) Which of the payoff matrices describes a complete market?
(b) In the case of the second and the fourth financial market, is it arbitrage-free, i.e. does there exist any arbitrage opportunity?

4.13. Suppose there are three states, $s = 1, 2, 3$ and two assets: a risky asset delivering the returns $[10\%, 5\%, -5\%]$ in the three states $s = 1, 2, 3$, and a risk-free asset with return equal to 2% in all states.

(a) Is the financial market complete? Give a return vector that cannot be hedged by the two existing assets.
(b) Determine the set of risk adjusted probabilities for which the expected return of the risky asset equals the risk-free rate.

Next, introduce a third asset with the return vector $[1\%, 2\%, 0\%]$.

(c) Find risk adjusted probabilities such that for both risky assets their expected return equals the risk-free rate. Is the market including the third asset arbitrage-free? [Hint: the answer may depend on the qualitative properties of the utility function!]

4.14. There are two assets and two states. Let the vector of expected returns be $\mu = (20\%, 10\%)$ and the covariance matrix

$$\text{cov} = \begin{pmatrix} 30\% & \sqrt{0.2}\sqrt{0.3}\cdot 100\% \\ \sqrt{0.2}\sqrt{0.3}\cdot 100\% & 20\% \end{pmatrix}.$$

Assume $\text{prob}_s = 0.5$. Find R such that $\mu(R^k) = \sum_{s=1}^{S} \text{prob}_s R^k_s = \text{prob}' R^k$ and

$$\begin{aligned} \text{cov}(R) &= \begin{pmatrix} \text{cov}(R^1, R^1) & \cdots & \text{cov}(R^1, R^K) \\ \vdots & & \vdots \\ \text{cov}(R^K, R^1) & \cdots & \text{cov}(R^K, R^K) \end{pmatrix} \\ &= R' \begin{pmatrix} \text{prob}_1 & \cdots & 0 \\ \vdots & \ddots & \vdots \\ 0 & \cdots & \text{prob}_S \end{pmatrix} R - \mu(R)\mu(R)'. \end{aligned}$$

4.15 (Raiffeisen Interest Notes Product). There are two assets and two equally likely states. The returns matrix is

$$R = \begin{pmatrix} \mu_1 + \rho\sigma_1 & \mu_2 - \sigma_2 \\ \mu_1 - \rho\sigma_1 & \mu_2 + \sigma_2 \end{pmatrix},$$

where $\sigma_1, \sigma_2 > 0$, $\mu_1, \mu_2 > 1\%$ and $\rho > 0$.

(a) Verify that $\mathbb{E}(R^k) = \mu_k$, $k = 1, 2$, and that the correlation between the two asset returns is -1.
(b) A structured product delivers the returns

$$\hat{R}^p = \max\left(\frac{\hat{R}^1 + \hat{R}^2}{2}, 1\%\right), \quad \text{where } \hat{R}^k = \begin{cases} N & \text{if } R^k > 1\%, \\ R^k & \text{if } R^k \leq 1\%. \end{cases}$$

Compute $\hat{R}^p$ and the likelihood of getting the return N as a function of the parameters $\mu_1, \mu_2, \sigma_1, \sigma_2$, and ρ. Assume $N > 1\%$.

4.16. Consider two assets (stocks and bonds) and two factors (oil price and growth rate) each of which can be high, h, or low, l, for simplicity for a total of four states.

Suppose the returns in those states are given by the first two matrices and the joint probabilities of the factor combinations are given by the second matrix:

Stocks	Oil h	Oil l	*Bonds*	Oil h	Oil l	*Prob*	Oil h	Oil l
Growth h	+1%	+5%	Growth h	−3%	−1%	Growth h	5%	30%
Growth l	−2%	−3%	Growth l	−1%	+2%	Growth l	50%	15%

(a) Calculate the state-space matrix.
(b) Determine a non-trivial factor loadings decomposition. Choose the factor values arbitrarily.
(c) Calculate the joint distribution of asset returns.

4.17. Consider four states and two factor return vectors in those states: growth boom $\approx$ $(3\%, 2\%, -1\%, 2\%)$ and inflation $\approx$ $(-2\%, -1\%, 2\%, 4\%)$. Suppose there are two assets (stocks and bonds) with the following factor loadings:

β_f^k	Stocks	Bonds
Inflation	−3	1
Growth	1	−2

(a) Determine the state-space matrix.
(b) Calculate the joint distribution of asset returns, supposing states are equally likely.

4.18. Consider the following CAPM economy: there are three equal likely states. The returns of the market portfolio in the three states are 6%, 2% and −2%. An investor holds a portfolio, which pays 23%, −4% and −1% in the three states. This investor has strictly monotone preferences and no background risk. The risk-free rate is zero.

(a) Check that the investors portfolio is feasible, i.e. the portfolio should lie on the SML.
(b) Is the portfolio of the investor optimal? Explain your answer.
(c) If the portfolio is not optimal, find a better portfolio which is feasible.
(d) What is the weakest possible assumption that can be made on the preferences of the investor such that your improvement still works? Do they have to be strictly monotone?

4.19. Assume a market with K normally distributed assets. The density function of a normal distribution is

$$f(x; \mu, \sigma^2) = \frac{1}{\sqrt{2\pi\sigma^2}} \exp\left(-\frac{1}{2}\frac{(x-\mu)^2}{\sigma^2}\right).$$

(a) Show that under the previous assumption any expected utility function $\mathbb{E}[u(X)]$ can be represented by a mean-variance utility function $MV(\mu, \sigma)$.
(b) Under which conditions to u is MV increasing in μ and decreasing in σ?
(c) Show that in the case of a Cumulative Prospect Theory-utility maximizer, with the probability weighting function w, the utility is still increasing in μ.

4.20. In an economy with two periods, two equally likely states and two assets, there is a representative agent with expected ln-utility endowed with the risky asset. There is no consumption in the first period. The risk-free asset has an interest rate of R. The risky asset pays in one state $S_u = D(\mu + \sigma)$ and in the other state $S_d = D(\mu - \sigma)$.

(a) Determine the price S_0 of the risky asset in the first period.
(b) Show that S_0 is decreasing in σ. σ can be seen as the volatility of the dividends of the stock.
(c) Calculate the volatility and the expected value of the returns of the risky asset.
(d) Show that the price of the risky asset decreases if the volatility of its returns increase.

4.21. Consider a market with two assets and three states, each with a probability of $\frac{1}{3}$. The prices of the two assets are denoted by $q_1 = 1$(normalized) and q_2 and the payoff-matrix given by matrix

$$\mathbf{A} = \begin{pmatrix} 1 & 0 \\ 0 & 1 \\ 0 & 1 \end{pmatrix}.$$

The utility of the agents are described by the logarithmic expected utility and the agents have heterogeneous beliefs, i.e., one agent maximizes the consumption in state 1 and 2 and the other agent in 1 and 3. The initial endowment are given by $w^1 = \left(1, \frac{1}{3}, 5\right)'$ and $w^2 = \left(1, 5, \frac{1}{3}\right)'$.

(a) Show that $q_2 = 3$ is an equilibrium price.
(b) Describe the risk-neutral probabilities in equilibrium.
(c) Show that every likelihood ratio process (as function of the aggregate wealth) has an increasing part.

4.22. In a two-period economy with three states, the payoff matrix A and the price vector are given as

$$A = \begin{pmatrix} 1 & 4 \\ 1 & 2 \\ 1 & 0 \end{pmatrix} \qquad q = \begin{pmatrix} 1 \\ 2 \end{pmatrix}.$$

(a) Is this financial market complete?
(b) Is the market arbitrage-free?
(c) A representative investor has an initial endowment of one of both assets and a utility function $U = \ln(c_1) + \ln(c_2) + \gamma \ln(c_3)$. Find a γ such that in a pure exchange economy with no first period consumption the q from above is the equilibrium price vector.

4.23. Consider a two period exchange economy with one representative investor and two equally likely states (an up and a down state). There are two assets: a riskless asset with an interest rate of $R = 1 + r$ (independent of the states) and a stock with a final payoff of $\mu + \sigma$ in state u and $\mu - \sigma$ in the state d. r, μ and σ are strictly positive. The riskless asset is in zero net supply and the representative investor gets one unit of stock as initial endowment. There is no first period consumption.

Let the utility function of the representative investor be

$$U^R = \mathbb{E}\left((c-1) - \frac{\gamma}{2}(c-1)^2\right) \quad \text{with } \gamma > 0.$$

(a) Are the preferences implied by U^R strongly monotonic? Why or why not?
(b) Express U^R in terms of a constant, the expected value of c and the variance of c.
(c) Determine the equilibrium stock price S. Write S as a function of γ, μ, σ and R.

4.24.

(a) Consider an economy with two time periods but without uncertainty. There are two agents with exogenous income of 1 in both periods. The agents' utility functions are given by:

$$U^1\left(c_0^1, c_1^1\right) = \ln(c_0^1) + 2\ln(c_1^1), \qquad U^2\left(c_0^2, c_1^2\right) = 2\ln(c_0^2) + \ln(c_1^2).$$

Determine the competitive equilibrium.

(b) Consider now an economy with two time periods and with uncertainty. Ignore consumption in the first period. There are two agents with exogenous income of $(1, 0)$ and $(0, 1)$ respectively in the two states of the second period. There are no endowments in the first period. The agents utility functions are given by:

$$U^1\left(c_1^1, c_2^1\right) = \ln(c_1^1) + \ln(c_2^1), \qquad U^2\left(c_1^2, c_2^2\right) = \ln(c_1^2) + \ln(c_2^2).$$

Determine the competitive equilibrium.

(c) Consider now an economy with two time periods and with uncertainty. Ignore consumption in the first period. There are two agents with exogenous income of $(1, 1)$ and $(1, 1)$ respectively in the two states of the second period. There are no endowments in the first period. The agents utility functions are given by:

$$U^1\left(c_1^1, c_2^1\right) = 2\ln(c_1^1) + \ln(c_2^1), \qquad U^2\left(c_1^2, c_2^2\right) = \ln(c_1^2) + 2\ln(c_2^2).$$

Determine the competitive equilibrium.

(d) Which motives of trade are present in the cases (a)–(c) given above?

4.25 (FTAP). Let q be the asset prices and A the payoff of the assets in a two-period economy with S states. There are no short sale restrictions or any other frictions. Now prove that the existence of strictly positive state prices, π, such that asset prices, q, are equal to the state-price weighted second period payoffs, rules out arbitrage opportunities.

4.26. Take the data for this exercise from the homepage of this book (you find it at http://www.financial-economics.de). There are 37 observations of index prices and risk-free rates. To prepare the data, calculate the gross returns for the 36 given years. Then, assume that every point of time represents one state of the economy with equal probabilities and show that there is no state-preference arbitrage. Use the mean of the risk-free rates given as the return of the risk-free asset. Include the World Index in the set of risky assets. To show no-arbitrage, you will use the FTAP implication that says it is equivalent to the existence of normalized state prices π_s^* with

(a) $\forall s,\ \pi_s^* > 0$
(This implies no-arbitrage for strict monotonic utility)
(b) $\forall s,\ \pi_s^* \geq 0$ and $\exists s',\ \pi_{s'}^* > 0$
(This implies no-arbitrage for weak monotonic utility)
(c) $\sum_s \pi_s^* > 0$
(This implies no-arbitrage for mean-variance utility).

4.27. In an economy with two states, the price of a stock in the first period is S and in the second period the stock pays uS in the up state and dS in the down state ($u > 1$ and $d < 1$). In the first period, a bond costs B and in the second period the bond pays RB independent of the state.

(a) Determine the value of a call option which pays C_u in the up state and C_d in the down state. Do this via the hedge portfolio.
(b) Do the same via the state prices (or the risk neutral measure).
(c) Determine the value of a put option with $S = 100$, $u = 2$, $d = 0.5$, $R = 1.1$ and a strike price $K = 100$.

4.28. In an economy with four states we have a bond and a stock:

$$A = \begin{pmatrix} 1 & 0.1 \\ 1 & 0.9 \\ 1 & 1.2 \\ 1 & 3.0 \end{pmatrix} \qquad q = \begin{pmatrix} 0.9 \\ 1.0 \end{pmatrix}.$$

There is also a barrier option with $A_3 = (0.5, 0.9, 1.2, 1.5)'$.

(a) Is it possible to hedge the barrier option by the stock and the bond?
(b) Find lower and upper price bounds of the barrier option. Do this numerically with, for example, Excel.
(c) Assume that all states are equally likely and the price of the barrier option is 0.95. There is an expected utility maximizer with $u(x) = x^\alpha$ and a prospect utility maximizer with

$$v_{KT}(R) = \begin{cases} (R - RP)^{\alpha^+} & \text{if } R > RP \\ -\beta\,(RP - R)^{\alpha^-} & \text{if } R \le RP \end{cases}$$

and $w(p) = (0.3, 0.2, 0.2, 0.3)'$. Assume that the investors put their whole wealth in the stock, the bond or the barrier option (only in one of these three assets). The initial wealth of the investors is 1.

Find preference parameters such that the prospect utility maximizer prefers the barrier option over the bond and the stock and the expected utility maximizer prefers the stock over the other two asset classes. Do this numerically with, for example, Excel.

4.29. In an economy with two assets and three states we have:

$$A = \begin{pmatrix} 1 & 1 \\ 1 & 0 \\ 1 & 0 \end{pmatrix} \qquad q = \begin{pmatrix} 0.90 \\ 0.25 \end{pmatrix}$$

(a) Determine the upper and the lower price bound of a third asset with $A_3 = (1, 2, 0)'$. Use the state price approach.
(b) Determine in the original market the upper and the lower price bound of $A_4 = (1, 0, 1)'$. Use the hedge-portfolio approach.

4.30 (FTAP). Suppose we have non-negative payoffs and short sales constraints, i.e. $A_s^k \ge 0$ and $\theta_k^i \ge 0$. Prove that the Fundamental Theorem of Asset Pricing reduces to: there is no $\theta \ge 0$ such that $q\theta \le 0$ and $A\theta > 0$ is equivalent to $q \gg 0$.

4.31. Consider an economy with two states: a boom, $s = 1$, and a recession, $s = 2$. The probability p of the boom is commonly known to the agents. We assume $p = \frac{5}{7}$.

There are two assets with the payoff matrix

$$A = \begin{pmatrix} 0.5 & 2 \\ 1 & 0 \end{pmatrix}.$$

The agents have logarithmic expected utility functions given by

$$U^i\left(c_1^i, c_2^i\right) := p \log\left(c_1^i\right) + (1-p)\log\left(c_2^i\right),$$

where c_s^i denotes consumption of agent $i \in \{1, 2\}$ in state $s \in \{1, 2\}$. The first (second) agent owns one unit of the first (second) asset, i.e., $\theta_A^1 = (1, 0)$ and $\theta_A{}^2 = (0, 1)$. There are no other endowments.

(a) Is the financial market complete?
(b) Compute the equilibrium consumption allocation and the state prices.
(c) Compute the equilibrium asset allocation θ and the asset prices.
(d) Compute the equilibrium asset allocation λ and the asset returns.

Suppose the returns are driven by two factors, $f = 1$, inflation, and $f = 2$, growth. The factor returns are given by $F = \begin{pmatrix} -1 & 1 \\ 0.5 & 0.5 \end{pmatrix}$.

(e) Compute the factor loadings β.
(f) Compute the equilibrium allocation of factors and the factor prices.

4.32. Consider an economy with two time periods, but without uncertainty. There are I agents with exogenous income of $w_1^i = (1+g)w_0^i$. The asset market consists of the risk-free asset. The agents' utility functions are given by:

$$U^i(c_0^i, c_1^i) = \ln(c_0^i) + \frac{1}{1+\delta}\ln(c_1^i)$$

Determine the real rate of interest as a function of the time preference and the growth rate of endowments.

4.33. Consider an economy with two periods and two states in the second period. The upper state has a probability p and the lower state a probability of $1 - p$. The initial endowment of the representative investor is a stock which has a return of u in the upper state and d in the lower state. The value of one stock is 1 in the first period. The utility function of the investor is

$$U(c_u, c_d) = p\log(c_u) + (1-p)\log(c_d).$$

The stock and a risk-free asset with zero net supply can be traded on the market.

(a) Given the gross risk-free rate R_f, determine risk-neutral probabilities π^* from the no arbitrage condition.
(b) Express π^* only in exogenous variables (i.e. without R^f). Use the equilibrium model for that.
(c) Determine R^f from the equilibrium model and the no arbitrage condition.
(d) Determine the risk-free rate R^f and the risk-neutral probabilities for

$$u(c) = \frac{c^{1-\gamma}}{1-\gamma}.$$

4.34. Assume a two period economy with a representative investor, S states in the second period, and the risk-free rate R_f. The utility function of the representative agent is

$$U = u(c_0) + \delta \sum_{s=1}^{S} p_s u(c_s)$$

The initial endowment of the agent is w_0 in the first period and w_s in the second period. Determine the likelihood ratio process in dependence of c_s for the following utility functions:

(a) $u(c) = c - \frac{\gamma}{2}c^2$ (quadratic)
(b) $u(c) = \frac{c^{1-\rho}}{1-\rho}$ (constant relative risk aversion, CRRA)
(c) $u(c) = -e^{-\alpha c}$ (constant absolute risk aversion, CARA)
(d)

$$u(c) = \begin{cases} (c - RP)^{\alpha^+}, & \text{if } c > RP, \\ -\beta\,(RP - c)^{\alpha^-}, & \text{if } c \leq RP, \end{cases}$$

(e)

$$u(c) = \begin{cases} (c - RP) - \alpha^+\,(c - RP)^2, & \text{if } c > RP, \\ -\beta\left((RP - c) - \alpha^-\,(RP - c)^2\right), & \text{if } c \leq RP. \end{cases}$$

4.35. Take the prepared data used in Exercise 4.26 and use the World Index as well as the risk-free returns. Assume that every point in time represents one state of the economy with equal probabilities. Calculate (normalized) state prices such that the quadratic distance to the (normalized) state prices of the CAPM is as small as possible and no arbitrage holds. For this you will need to use FTAP and the normalized state prices expression implied by CAPM. To show this, report the computed state prices being positive and that the quadratic distance to normalized state prices is very close to zero.

4.36. We are in the same setup as in Exercise 4.34 and $\delta = \frac{1}{1+r_f}$. Assume that the likelihood ratio process depends linearly on several factors: $l = 1 + b'(f - \mathbb{E}(f))$. The excess returns are defined as $R^{e,k} = R^k - R_f$.

(a) How can the expression $\beta^k = \text{var}(f)^{-1} \text{cov}(f, R^{e,k})$ be interpreted?
(b) Show that $\mathbb{E}(R^{e,k}) = \lambda' \beta^k$, where $\lambda = -\text{var}(f)b$.
(c) How can λ and b be estimated from data?

4.37. A pension fund is well diversified and the actual beta of its portfolio is 0.5. The CAPM risk premium $\mathbb{E}\left(R_M - R_f\right)$ is 5% and the risk-free rate is 3%. The board of the pension fund wants to take into account the risk of a recession and the inflation risk. They want to do this with a two factor APT-model. They take the unexpected inflation and the unexpected changes of GDP as risk factors:

$$\mathbb{E}(R_i) - R_f = b^i_{Inf}\left(\mathbb{E}(R_{Inf}) - R_f\right) + b^i_{GDB}\left(\mathbb{E}(R_{GDP}) - R_f\right)$$

They estimate the model and obtain:

$$\mathbb{E}(R_i) = 3\% - b^i_{Inf} \cdot 1\% + b^i_{GDP} \cdot 2\%$$

(a) The sensitivity of the current portfolio to an unexpected change of the GDP is 0.2. What is the sensitivity to the unexpected inflation of the current portfolio?
(b) The sensitivity to unexpected changes of the GDP and the inflation should be reduced to zero. What is the expected return of such a portfolio?
(c) The board of the pension fund has decided that they want an average return of 6% and no exposure to the unexpected inflation. What is the sensitivity of that portfolio to unexpected changes of the GDP?

4.38. Consider a two-period economy, $t = \{0, 1\}$, with two possible states in the second period $s = \{1, 2\}$. Assume that consumption only takes place in $t = 1$. There are two agents $i = \{1, 2\}$ having the logarithmic expected utilities

$$U^1(c_1, c_2) = 0.75 \ln(c_1) + 0.25 \ln(c_2),$$
$$U^2(c_1, c_2) = 0.25 \ln(c_1) + 0.75 \ln(c_2),$$

respectively. There are two assets in unit supply: one risk-free asset paying off 1 in both states and one risky asset paying off 2 in the first state and 0.5 in the second state. The first (second) agent owns one unit of the first (second) asset. Assets are the only source of income.

(a) Determine the competitive equilibrium.
(b) Find a representative consumer with logarithmic expected utility function whose demand could also generate the equilibrium prices found in (a).

(c) Suppose the payoff of the second asset increases to 3 in the first state. Compute the new asset prices using the representative consumer as determined in (b).
(d) Compute the new equilibrium prices in the original economy with two agents.

4.39 (Financial Innovation and Representative Agent). Consider a two-period financial economy without consumption in the first period. There are $k = 1, \ldots, K$ assets, $i = 1, \ldots, I$ consumers and $s = 1, \ldots, S$ states in the second period. The assets are of zero net supply. Consumer i gets an initial endowment w^i in the second period and no endowment in the first period.

(a) Define the financial equilibrium in this economy.

Assume that there are three states and two consumers such that

$$U^1(c_1, c_2, c_3) = \ln c_1 + \ln c_2, \qquad w^1 = (0, 1, 2)',$$

$$U^2(c_1, c_2, c_3) = \ln c_2 + \ln c_3, \qquad w^2 = (2, 1, 0)'.$$

(b) Show that for non-negative consumption plans and the payoff matrix

$$A = \begin{pmatrix} 1 & 0 \\ 0 & 1 \\ 0 & 1 \end{pmatrix},$$

$\theta^{1,*} = (2, -1/2)', \theta^{2,*} = (-2, 1/2)'$ and $q^* = (1, 4)'$ is an equilibrium.

Consider a third asset with the payoff $A^3 = (0, 0, 1)'$.

(c) Is it possible to duplicate this asset from the assets in the previous question?
(d) Calculate an arbitrage-free price for the third asset (out of A and q^*). Is it unique?
(e) Determine the equilibrium in the economy including the third asset. Assume non-negative consumption.
(f) Check that a representative consumer with $U(c) = \frac{1}{5}\ln(c_1) + \frac{1}{5}\ln(c_2) + \frac{3}{5}\ln(c_3)$ and the aggregated endowment $w = w^1 + w^2$ the case without financial innovation generates in the same asset prices as in (b).
(g) Show that this representative consumer would misprice the financial innovation.

5 Multiple-Periods Model

5.1. Let us consider the following three-period model where the returns of two assets are marked at each node:

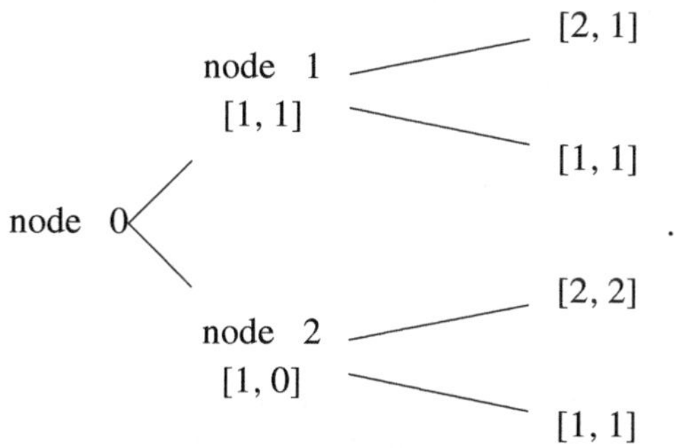

Is the market with these two assets complete?

5.2. Consider a T-period economy with i.i.d. relative dividends $D^k / \sum_j D^j$. The investors are all expected utility maximizers with homogeneous rational beliefs. The investor i has the utility

$$U^i(c^i) = \mathbb{E}_P \sum_{t=1}^{\infty} (\delta^i)^t \log(c_t^i).$$

T. Hens, M. O. Rieger, *Solutions to* Financial Economics,
Springer Texts in Business and Economics,
https://doi.org/10.1007/978-3-662-59889-4_5

Show that

$$\lambda_t^{i,k} = (1 - \lambda^{i,c})\mathbb{E}_P\left(\frac{D^k}{\sum_j D^j}\right)$$

is a perfect foresight equilibrium.

5.3. Assume an economy with three time periods $t = 0, 1, 2$ with two assets and a representative expected log-utility maximizer, who has no time discounting. In $t = 1$ there are two equally likely states u and d. If u (d) occurred, the economy switches with equal probabilities into the states uu (du) and ud (dd). One asset is a riskless bond, which pays 1 in every period (also in $t = 0$), and a risky stock, which pays 1 in $t = 0, 2$ in u and uu, 0.5 in d, ud, dd and 1.5 in du. The initial endowment of the representative investor, θ_{-1}, is one unit of the bond and one unit of the stock. There is no other source of income.

(a) Determine the stock price in $t = 2$
(b) Determine the equilibrium prices of the bond and the stock in $t = 0$ and in the two states in $t = 1$.
(c) Is the market complete?
(d) Determine the state prices.
(e) Is the market arbitrage-free?
(f) Calculate the arbitrage-free value of a European call in $t = 0$ with maturity $t = 2$. The strike price is 1 and the payoff of the call is in $t = 2$: (Dividend of the stock − strike price)$^+$.
(g) Calculate the arbitrage-free value of an American call with the same strike and the same payoff function. But additionally the option can be exercised at any point before. The payoff is then (stock price without dividends − strike price)$^+$.
(h) Calculate the risk-free rates in period 0 with maturity t: $R_{f,t} = 1 + r_{f,t}$.
(i) Check that the following relations hold for the stock between $t = 0$ and $t = 1$:
  (i) $q_t^k = \frac{1}{1+r_{f,t}}\mathbb{E}_{\pi_t^*}\left(D_{t+1}^k + q_{t+1}^k\right)$
  (ii) $\mathbb{E}_{\pi_t^*}\left(R_{t+1}^k\right) = R_{f,t}$
  (iii) $\mathbb{E}_{P_t}\left(R_{t+1}^k\right) = R_{f,t} - \mathrm{cov}_{P_t}\left(R_{t+1}^k, l_{t+1}\right)$ with $l = \frac{\pi^*}{P}$

5.4. The actual interest rate of a risk-free zero coupon bond over 5 years is 3% and for a time period of 8 years it is 5%. Determine the forward rate $f(0, 5, 8)$ by no arbitrage.

5.5. There is a three period economy with $t = 0, 1, 2$ and a representative investor with a utility function:

$$U(c) = \ln(c_0) + \frac{1}{1+\delta}\mathbb{E}(\ln(c_1)) + \frac{1}{(1+\delta)^2}\mathbb{E}(\ln(c_2)).$$

The agent can only trade in $t = 0$. He can only trade either bonds with a time to maturity of 1 and 2 or the forward $f(0, 1, 2)$.

(a) Assume that there is no uncertainty in the model, i.e. the initial endowment is w_0, w_1 and w_2. Furthermore the prices of the consumption goods are p_0, p_1 and p_2. Determine the term structure in and the forward rate at $t = 0$ and the realized spot rate in $t = 1$.

(b) Assume now that the investor discounts hyperbolically, i.e.

$$U^H(c) = \ln(c_0) + \frac{1}{1+\beta}\left(\frac{1}{1+\delta}\mathbb{E}\left(\ln(c_1)\right) + \frac{1}{(1+\delta)^2}\mathbb{E}\left(\ln(c_2)\right)\right).$$

Determine the term structure in and the forward rate at $t = 0$ and the realized spot rate in $t = 1$.

(c) In the case with the standard time discounting uncertainty is added to the model. The initial endowment and the prices in $t = 0$ remain unchanged. But in $t = 1$ there is a liquidity shock and the economy switches into an upper state with probability q and a lower state with probability $1 - q$. In $t = 1, 2$ the initial endowment and prices of the consumption good depend on the state occurred.

(i) Determine the term structure and the forward rate at $t = 0$. Find also the expected and the realized spot rates in $t = 1$.

(ii) The nominal growth rate is defined as

$$\overline{g}_{t,t+1}(s) = \frac{w_{t+1}(s)p_{t+1}(s)}{w_t(s)p_t(s)} - 1.$$

Calculate the term structure, the forward rate and the realized spot rate in the first period, given that $\delta = 0.1$, $q = 0.5$, $\overline{g}_{0,1,u}(u) = \overline{g}_{1,2,u} = \frac{1}{9}$, $\overline{g}_{0,1,d} = -\frac{1}{21}$ and $\overline{g}_{0,2,d} = 0$.

(d) The observed term structures are typically increasing and there is a forward rate bias. Which of those features can be explained by the models above?

5.6. There are two assets and two states which are equally likely. The following is the dividend matrix:

$$\begin{bmatrix} 3 & 1 \\ 1 & 5 \end{bmatrix},$$

where each column represent the possible payoffs of one asset.

Calculate the evolutionary portfolio strategy λ^*.

5.7. Consider the exponential growth rate in the evolutionary portfolio model with i.i.d. dividends and time independent strategies.

$$g(\hat{\lambda}, \hat{\lambda}^M) = \mathbb{E}_p \ln \left(1 - \lambda^c + \lambda^c \sum_{k=1}^{K} \frac{d^k \hat{\lambda}^k}{\hat{\lambda}^{M,k}}\right).$$

(a) Show that $g(\hat{\lambda}^M, \hat{\lambda}^M) = 0$.
(b) Show that $g(\hat{\lambda}^{1/k}, \hat{\lambda}) > 0$ if $\hat{\lambda}^k = \varepsilon$ for some k and $\varepsilon > 0$.
(c) Show that $\hat{\lambda}^* = \mathbb{E}_P d^k$ maximizes $g(\hat{\lambda}, \hat{\lambda}^*)$ over all $\hat{\lambda}$.

6 Theory of the Firm

6.1 (Production Technology Set). Suppose $S = 1$ and wealth in period one is produced from wealth in period zero by the production function $\sqrt{}$. Define the production technology set $Y \subset \mathbb{R}^2$ and check whether the properties (i)–(v) given in Assumption 6.1 in the book are satisfied.

6.2 (Dividend Policy). In extension of Exercise 6.1 suppose there are three states $s = 0, 1, 2$. Moreover, the production decision is $y^j = \begin{pmatrix} -1 \\ 1.5 \end{pmatrix}$ and the asset payoff matrix is $A = \begin{pmatrix} 1 & 1 \\ 1 & 0 \end{pmatrix}$ while asset prices are $q = (2, 1)$. Determine the three financial policies ξ^j of

(a) 100% equity finance,
(b) 100% bond finance,
(c) 100% risk coverage.

6.3 ("Undo" Portfolio). Suppose there are three states of the world $s = 0, 1, 2$ and two firms $j = 1, 2$. A household i has chosen the portfolio $z^i = (1, 1)^T$ in a situation where the two firms use the financial policy

$$\xi^1 = \begin{pmatrix} -1 \\ 1 \end{pmatrix}, \xi^2 = \begin{pmatrix} 1 \\ -1 \end{pmatrix}.$$

The household owns 1% of shares in both firms, i.e. $\bar{\delta}^i_j = \frac{1}{100}$, $j = 1, 2$. Compute the "undo" portfolio $\hat{z}^i$ undoing a change in financial policy

$$\hat{\xi}^1 = \begin{pmatrix} 1 \\ 0 \end{pmatrix}, \hat{\xi}^2 = \begin{pmatrix} -1 \\ 1 \end{pmatrix}.$$

T. Hens, M. O. Rieger, *Solutions to* Financial Economics,
Springer Texts in Business and Economics,
https://doi.org/10.1007/978-3-662-59889-4_6

6.4 (Bankruptcy and Limited Liability).

(a) Suppose there are three states $s = 0, 1, 2$. The payoff matrix is

$$A = \begin{pmatrix} 1 \\ 1 \end{pmatrix}$$

(there is only a risk-free asset). Assuming unlimited liability as in Eq. (6.3) of the book, show that the financial policy ξ^j does not change the set of attainable wealth transfers $\langle A, D \rangle = \{x \in \mathbb{R}^2 \mid x = Az + D^j \delta, \text{for some } z, \delta\}$.

(b) Now assume limited liability i.e.

$$d_s^j = max\{0, y^j + \begin{pmatrix} -q \\ A \end{pmatrix}\}, s = 0, 1, 2.$$

Give an example of two financial policies ξ^j leading to different $\langle A, D \rangle$.

6.5 (Fisher Separation Theorem). Suppose $s = 0, 1$ (no uncertainty) and there is a single firm $j = 1$ owned equally by two consumers $\bar{\delta}_j^1 = \bar{\delta}_j^2 = \frac{1}{2}$. The firm's production technology is

$$y^j = \{(y_0^j, y_1^j) \mid y_1^j \leq \sqrt{-y_0^j}\}.$$

Suppose the first consumer is rich today, while the second is rich tomorrow, i.e. $w^1 = (1, 0) = w^2$. The consumers also differ with respect to their (true) preferences. Consumer 1's utility is $U^1(x_0^1, x_1^1) = \ln x_0^1 + 2 \ln x_1^1$, while consumer 2's utility is $U^2(x_0^2, x_1^2) = 2 \ln x_0^2 + \ln x_1^2$. Note that the firm's decision rule is not well defined since there are no asset and state prices π^j are unrestricted.

(a) What is the firms best production plan from the point of view of the two owners?

(b) Suppose there is a risk-free asset $A = \begin{pmatrix} 1 \\ 1 \end{pmatrix}$. Compute the financial equilibrium with production according to Definition 6.3 in the book. for the case of non-incorporated companies and show that both owners agree on the chosen production plan.

6.6 (Drèze Criterion). Reconsider part (a) of Exercise 6.5 and introduce the Drèze Criterion as the decision rule of the firm. Compute the corresponding Drèze-equilibrium (i.e. the financial market equilibrium with the Drèze rule as the decision criterion of the firm).

6.7 (Drèze Equilibria with Incorporated Companies). Assume the following model of incorporated companies

$$\mathrm{GEI} = \left(\mathbb{R}^{S+1}, (U^i, \omega^i, \delta^i)_{i=1,\dots,I}, (Y^k)_{k=1,\dots,K}\right)$$

with $I = S = K = 2$ (Note: there is no stock market A) and the following specifications

$$U^1(x_0, x_1, x_2) = 3x_1 + x_2\,,\; U^2(x_0, x_1, x_2) = x_1 + 3x_2\,,\; \omega^1 = \begin{pmatrix}1\\0\\0\end{pmatrix} = \omega^2.$$

The technology sets are

$$Y^1 := \left\{ y \in \mathbb{R}^3 \mid y = \alpha^1 \begin{pmatrix}-1\\2\\0\end{pmatrix} + \beta^1 \begin{pmatrix}-1\\0\\1\end{pmatrix},\; \text{for } \alpha^1, \beta^1 \geq 0 \right\},$$

$$Y^2 := \left\{ y \in \mathbb{R}^3 \mid y = \alpha^2 \begin{pmatrix}-1\\1\\0\end{pmatrix} + \beta^2 \begin{pmatrix}-1\\0\\2\end{pmatrix},\; \text{for } \alpha^2, \beta^2 \geq 0 \right\}.$$

(a) Show the supply $(Y^i + \omega^i)$ in a diagram for the output in $t = 1$. Also draw the consumers' indifference curves.

(b) Show that there are the following two Drèze equilibria

(i) $\overset{*}{x}{}^1 = \begin{pmatrix}0\\1\\0\end{pmatrix},\; \overset{*}{x}{}^2 = \begin{pmatrix}0\\0\\1\end{pmatrix},\; \overset{*}{y}{}^1 = \begin{pmatrix}-1\\0\\1\end{pmatrix},\; \overset{*}{y}{}^2 = \begin{pmatrix}-1\\1\\0\end{pmatrix},$

$\overset{*}{z}{}^1 = \begin{pmatrix}0\\1\end{pmatrix},\; \overset{*}{z}{}^2 = \begin{pmatrix}1\\0\end{pmatrix},\; \overset{*}{r} = 0.$

(ii) $\overset{*}{x}{}^1 = \begin{pmatrix}0\\2\\0\end{pmatrix},\; \overset{*}{x}{}^2 = \begin{pmatrix}0\\0\\2\end{pmatrix},\; \overset{*}{y}{}^1 = \begin{pmatrix}-1\\2\\0\end{pmatrix},\; \overset{*}{y}{}^2 = \begin{pmatrix}-1\\0\\2\end{pmatrix},$

$\overset{*}{z}{}^1 = \begin{pmatrix}1\\0\end{pmatrix},\; \overset{*}{z}{}^2 = \begin{pmatrix}0\\1\end{pmatrix},\; \overset{*}{r} = 0.$

(c) From the results obtained in (b), what can be concluded about the Pareto-efficiency of Drèze equilibria?

6.8. Suppose the Financial Market Equilibrium (FME) with Incorporated Companies is defined as follows:

A FME for a financial economy ε_F^P with incorporated companies consists of allocations $(\overset{*}{\phi}_X, \overset{*}{\phi}_Z, \overset{*}{\phi}_H, \overset{*}{\phi}_y, \overset{*}{\phi}_P, \overset{*}{\phi}_\tau)$ such that

(i) $(\overset{*}{x^i}, \overset{*}{z^i}, \overset{*}{\theta^i}) \in \mathbb{B}^i(\overset{*}{q}, \overset{*}{p}, \omega^i, \overline{\delta^i}, A, \overset{*}{D})$ and $\overset{*}{x_i} \in \arg\max U^i(x^i)$ for all $i \in I$,

(ii) $(\overset{*}{d^j}, \overset{*}{y^j}, \overset{*}{\zeta^j}, \overset{*}{\tau^j}) \in \mathbb{W}^i(\overset{*}{q}, \overset{*}{p}, A, \overset{*}{D})$ and $\overset{*}{d^j} \in \arg\max \pi^{j'} d^j$ for all $j \in J$,

(iii) $\sum_{i \in I} \overset{*}{x^i} \leq \sum_{j \in J} \omega^i + \sum_{j \in J} \overset{*}{y^j}$,

(iv) $\sum_{i \in I} \overset{*}{z^i} + \sum_{j \in J} \overset{*}{\zeta^j} = 0$, and $\sum_{i \in I} \overset{*}{\delta^i} + \sum_{j \in J} \overset{*}{\tau^j} = \mathbb{1}$, where π^j satisfies the NAC for all j.

Prove the irrelevance of the financial policy of the firm for the case of incorporated companies, stated by the Modigliani–Miller theorem:

Let $(\overset{*}{\phi_X}, \overset{*}{\phi_Z}, \overset{*}{\phi_H}, \overset{*}{\phi_y}, \overset{*}{\phi_P}, \overset{*}{\phi_\tau})$ and $(\overset{*}{q}, \overset{*}{p}, \{\pi^j\}_{j \in J})$ be a FME with $(I - \hat{\tau})$ invertible. Then there are $(\hat{\phi_Z}, \hat{\phi_H})$ such that

$$(\overset{*}{\phi_X}, \overset{*}{\phi_Z}, \overset{*}{\phi_H}, \overset{*}{\phi_y}, \overset{*}{\phi_P}, \overset{*}{\phi_\tau}),\ (\overset{*}{q}, \overset{*}{p}, \{\pi^j\}_{j \in J})$$

is an FME. The theorem suggests the interpretation that the financial policy of the firms is irrelevant to the stock prices $\overset{*}{p}$.

7 Information Asymmetries on Financial Markets

7.1. There are two time periods $t = 0, 1$ and two states in the second period $s = 1, 2$. There are two consumers $i = 1, 2$. The first consumer is rich today and poor tomorrow, $w^1 = (1, 0, 0)$. The second is rich tomorrow and poor today, $w^2 = (0, 1, 1)$. There are two Arrow securities, i.e. $A = \begin{pmatrix} 1 & 0 \\ 0 & 1 \end{pmatrix}$. The first consumer does not know which state occurs and a priori assigns equal probabilities to them. Before trading the asset the second consumer gets a signal revealing the state of the world. Both consumers have ln-utility of wealth, and no time discount rate. All this information is common knowledge. No consumer acts strategically (as if there were two types of infinitely many consumers).

(a) Compute the competitive equilibrium assuming the first consumer is naive, i.e. he does not try to infer the information of the second consumer from market prices.
(b) Compute the competitive equilibrium for the case that the signal to the second consumer does not reveal the state and he also beliefs they are equally likely.
(c) Now compute the competitive equilibrium in which the first consumer tries to infer the information of the second consumer from market prices.

7.2 (Extracted from [MCWG$^+$95], Exercise 13.B.4). Suppose two individuals, 1 and 2, are considering a trade at price p of an asset that they both use only as a store of wealth. Ms. 1 is currently the owner. Each individual i has a privately observed signal of the asset's worth y_i. In addition, each cares only about the expected value of the asset 1 year from now. Assume that a trade at price p takes place only if both parties think they are being made strictly better off. Prove that the probability of trade occurring is zero. [Hint: study the following trading game: the two individuals simultaneously say either "trade" or "no trade", and a trade at price p takes place only if they both say "trade."]

T. Hens, M. O. Rieger, *Solutions to* Financial Economics,
Springer Texts in Business and Economics,
https://doi.org/10.1007/978-3-662-59889-4_7

7.3 (Extracted from [Tir10], Exercise 3.12).

Consider a two date model of market finance with a representative consumer/investor. This consumer has utility of consumption $u(c_0)$ at date 0, the date at which he lends to the firm, and utility of consumption $u(c(\omega))$ at date 1, the date at which he receives the return from investment. There is macroeconomic uncertainty in that the representative consumer's date 1 consumption depends on the state of nature ω. The state of nature describes both what happens in this particular firm and in the rest of the economy (even though aggregate consumption is independent of the outcome in this particular firm to the extent that the firm is atomistic, which we will assume).

Suppose that the entrepreneur works. Let S denote the event "the project succeeds" and F the event "the project fails." Let

$$q_S = E\left[\frac{u'(c(\omega))}{u'(c_0)} \mid \omega \in S\right]$$

and

$$q_F = E\left[\frac{u'(c(\omega))}{u'(c_0)} \mid \omega \in F\right].$$

The firm's activity is said to covary positively with the economy (be "procyclical") if $q_S < q_F$, and negatively (be "countercyclical") if $q_F < q_S$.

Suppose that

$$p_H q_S + (1 - p_H) q_F = 1.$$

(a) Interpret this assumption.
(b) In the fixed-investment model (and still assuming that the entrepreneur is risk neutral), derive the necessary and sufficient condition for the project to receive financing.
(c) What is the optimal contract between the investors and the entrepreneur? Does it involve maximum punishment ($R_b = 0$) in the case of failure? How would your answer change if the entrepreneur were risk averse? (For simplicity, assume that her only claim is in the firm. She does not hold any of the market portfolio.)

7.4 (Extracted from [MCWG+95], Exercise 14.C.8). Air Shangri-la is the only airline allowed to fly between the islands of Shangri-la and Nirvana. There are two types of passengers, tourist and business. Business travelers are willing to pay more than tourists. The airline, however, cannot tell directly whether a ticket purchaser is a tourist or business traveler. The two types do differ, though, in how much they are willing to pay to avoid having to purchase their tickets in advance. (Passengers do not like to commit themselves in advance to traveling at a particular time.)

More specifically, the utility levels of each of the two types net of the price of the ticket P, for any given amount of time W prior to the flight that the ticket is purchased are given by

$$\text{Business:} v - \theta_B P - W,$$

$$\text{Tourist:} v - \theta_T P - W,$$

where $0 < \theta_B < \theta_T$. (Note that for any given level of W, the business traveler is willing to pay more for his ticket. Also, the business traveler is willing to pay more for any given reduction in W.) The proportion of travelers who are tourists λ. Assume that the cost of transporting a passenger is c. Assume in (a)–(d) that Air Shangri-la wants to carry both types of passengers.

(a) Draw the indifference curves of the two types in (P, W)-space. Draw the airline's isoprofit curves. Now formulate the optimal (profit-maximizing) price discrimination problem mathematically that Air Shangri-la would want to solve. [Hint: impose nonnegativity of prices as a constraint since, if it charged a negative price, it would sell an infinite number of tickets at this price.]
(b) Show that in the optimal solution, tourists are indifferent between buying a ticket and not going at all.
(c) Show that in the optimal solution, business travelers never buy their ticket prior to the flight and are just indifferent between doing this and buying when tourists buy.
(d) Describe fully the optimal price discrimination scheme under the assumption that they sell to both types. How does it depend on the underlying parameters λ, θ_B, θ_T, and c?
(e) Under what circumstances will Air Shangri-la choose to serve only business travelers?

7.5 (Extracted from [MCWG$^+$95], Exercise 14.C.9). Consider a risk-averse individual who is an expected utility maximizer with a Bernoulli utility function over wealth $u(.)$. The individual has initial wealth W and faces a probability θ of suffering a loss of size L, where $W > L > 0$. An insurance contract may be described by a pair (c_1, c_2), where c_1 is the amount of wealth the individual has in the event of no loss and c_2 is the amount the individual has if a loss is suffered. That is, in the event no loss occurs the individual pays the insurance company an amount $(W - c_1)$, whereas if a loss occurs the individual receives a payment $[c_2 - (W - L)]$ from the company.

(a) Suppose that individual's only source of insurance is a risk-neutral monopolist (i.e., the monopolsit seeks to maximize its expected profits). Characterize the contract the monopolist will offer the individual in the case in which the individual's probability of loss, θ, is observable.
(b) Suppose, instead, that θ is not observable by the insurance company (the individual knows θ). The parameter θ can take one of two values θ_L, θ_H, where $\theta_H > \theta_L > 0$ and $\text{Prob}(\theta_L) = \lambda$. Characterize the optimal contract offers of the

monopolist. Can one speak of one type of insured individual being "rationed" in his purchases of insurance (i.e., he would want to purchase more insurance if allowed to at fair odds)? Intuitively, why does this rationing occur? [Hint: it might be helpful to draw a picture in (c_1, c_2)-space. To do so, start by locating the individual's endowement point, that is, what he gets if he does not purchase any insurance.]

(c) Compare your solution in (b) with your answer to Exercise part (a).

7.6 (Extracted from [Tir10], Exercise 3.21). Analyze an entrepreneur's risk preferences with respect to net worth. The entrepreneur is risk neutral and protected by limited liability. The investors are risk neutral and demand a rate of return equal to 0. At date 0, the entrepreneur decides whether to insure against a date-1 income risk

$$r = A_0 + \varepsilon,$$

where $\varepsilon \in [\underline{\varepsilon}, \bar{\varepsilon}]$, $E[\varepsilon] = 0$, and $A_0 + \underline{\varepsilon} \geq 0$.

For simplicity, we allow only a choice between full hedging and no hedging (the theory extends straightforwardly to arbitrary degrees of hedging). Hedging (which wipes out the noise and thereby guarantees that the entrepreneur has cash on hand A_0 at date 1) is costless. After receiving income, the entrepreneur uses her cash to finance investment I and must borrow $I - A$ from investors, with A $=A_0$ in the case of hedging and $A = A_0 + \epsilon$ in the absence of hedging (provided that $A \leq I$; otherwise there is no need to borrow). Note that there is no overall liquidity management as there is no contract at date 0 with the financiers as to the future investment. This exercise investigates a variety of situations under which the entrepreneur may prefer either hedging or "gambling" (here defined as "no hedging").

(a) *Fixed investment, binary effort.* Suppose that the investment size is fixed, and that the entrepreneur at date 1, provided that she receives funding, either behaves (probability of success p_H, no private benefit) or misbehaves (probability of success p_L, private benefit B). As usual, the project is not viable if it induces misbehavior and has a positive NPV ($p_H R > I > p_L R + B$, where R is the profit in the case of success). Let $\bar{A}$ be defined by

$$p_H(R - \frac{B}{\Delta p}) = I - \bar{A}.$$

Suppose that ε has a wide support.

Show that the entrepreneur

- hedges if $A_0 \geq \bar{A}$
- gambles if $A_0 < \bar{A}$.

(b) *Fixed investment, continuous effort.* Suppose, that succeeding with probability p involves an unverifiable private cost $\frac{1}{2}p^2$ for the entrepreneur (so, effort in this subquestion involves a cost rather than the loss of a private benefit). (Assume R < 1 to ensure that probabilities are smaller than 1.)

Write the investors' breakeven condition as well as the NPV as functions of the entrepreneur's stake, R_b, in success. Note that one can focus without loss of generality on $R_b \in [\frac{1}{2}R,R]$. Assume that $I - A_0 < \frac{1}{4}R^2$ and that the support of ε is sufficiently small that the entrepreneur always receives funding when she does not hedge (and *a fortiori* when she hedges). This assumption eliminates the concerns about financing of investment that were crucial in question (a). Show that the entrepreneur hedges.

(c) *Variable investment.* Return to the binary effort case ($p = p_H$ or p_L), but assume that the investment I is variable. The income is RI in the case of success and 0 in the case of failure. The private benefit of misbehaving is $B(I)$ with $B' > 0$. Assume that the size of investment is always constrained by the pledgeable income and that the optimal contract induces good behavior.

Show that the entrepreneur

- hedges if $B(.)$ is konvex;
- is indifferent between hedging and gambling if $B(.)$ is linear;
- gambles if $B(.)$ is concave.

(d) *Variable investment and unobservable income.* Suppose that the investment size is variable and that the income from investment $R(I)$ is unobservable by investors (fully appropriated by the entrepreneur) and is concave. Suppose that it is always optimal for the entrepreneur to invest her cash on hand. Show that the entrepreneur hedges.

(e) *Liquidity and risk management.* Suppose, in contrast with Froot et al.'s analysis, that the entrepreneur can sign a contract with investors at date 0. Show that the entrepreneurs's utility can be maximized by insulating the date-1 volume of investment from realization of ε, i.e., with full hedging, even in situations where gambling was optimal when funding was secured only at date 1.

7.7 (Extracted from [MCWG+95], Exercise 13.C.6). Consider a market for loans to finance investment projects. All investment projects require an outlay of 1 dollar. There are two types of projects: good and bad. A good project has a probability of p_G of yielding profits of $\prod > 0$ and a probability $(1-p_G)$ of yielding profits of zero. For a bad project, the relative probabilities are p_B and $(1 - p_B)$, respectively $p_G > p_B$. The fraction of projects that are good is $\lambda \in (0, 1)$.

Entrepreneurs go to banks to borrow the cash to make the initial outlay (assume for now that they borrow the entire amount). A loan contract specifies an amount R that is supposed to be repaid to the bank. Entrepreneurs know the type of project they have, but the banks do not. In the event that a project yields profits of zero, the entrepreneur defaults on her loan contract, and the bank receives nothing. Banks are competitive and risk neutral. The risk-free rate of interest (the rate the banks pay to borrow funds) is r. Assume that

$$p_G \prod -(1+r) > 0 > p_B - (1+r).$$

(a) Find the equilibrium level of R and the set of projects financed. How does this depend on $p_G, p_B, \lambda, \prod$, and r?

(b) Now suppose that the entrepreneur can offer to contribute some fraction x of the 1 dollar initial outlay from her own funds ($x \in [0, 1]$). The entrepreneur is liquidity constrained, however, so that the effective cost of doing so is $(1+\rho)x$, where $\rho > r$.

(i) What is an entrepreneur's payoff as a function of her project type, her loan-repayment amount R, and her contribution x?

(ii) Describe the best (from a welfare perspective) separating perfect Bayesian equilibrium of a game in which the entrepreneur first makes an offer that specifies the level of x she is willing to put into a project, banks then respond by making offers specifying the level of R they would require, and finally the entrepreneur accepts a bank's offer or decides not to go ahead with the project. How does the amount contributed by entrepreneurs with good projects change with small changes in $p_B, p_G, \lambda, \prod$, and r?

(iii) How do the two types of entrepreneurs do in the separating equilibrium of (b) (ii) compared with the equilibrium (a)?

7.8 (Extracted from [MCWG+95], Exercise 13.D.2). Consider the following model of the insurance market. There are two types of individuals: high risk and low risk. Each starts with initial wealth W but has a chance that an accident (e.g., a fire) will reduce her wealth by L. The probability of this happening is p_L for low-risk types and p_H for high-risk types, where $p_H > p_L$. Both types are expected utility maximizers with a Bernoulli utility function over wealth of $u(w)$, with $u'(w) > 0$ and $u''(w) < 0$ at all w. There are two risk-neutral insurance companies. An insurance policy consists of a premium payment M made by the insured individual to her insurance firm and a payment R from the insurance company to the insured individual in the event of a loss.

(a) Suppose that individuals are prohibited from buying more than one insurance policy. Argue that a policy can be thought of as specifying the wealth levels of the insured individual in the two states "no loss" and "loss".

(b) Assume that the insurance companies simultaneously offer policies; they can each offer any finite number of policies. What are the subgame perfect Nash equilibrium outcomes of the model? Does an equilibrium necessarily exist?

7.9 (Extracted from [Tir10], Exercise 6.1). Consider a fixed-investment model and assume that only the borrower knows the private benefit associated with misbehavior. When the borrower has private information about this parameter, lenders are concerned that this private benefit might be high and induce the borrower to misbehave. In the parlance of information economics, the "bad types" are the types of borrower with high private benefit. We study the case of two possible levels of private benefit. The borrower wants to finance a fixed-size project costing I, and, for simplicity, has no equity (A = 0). The project yields R (success) or 0 (failure). The probability of success is p_H or p_L, depending on whether the borrower works or shirks, with $\Delta p \equiv p_H - p_L > 0$. There is no private benefit when working. The private benefit B enjoyed by the borrower when shirking is either $B_L > 0$ or $B_H > B_L$. The borrower will be labeled a "good borrower" when $B = B_L$ and a "bad borrower" when $B = B_H$. At the date of contracting, the borrower knows the level of her private benefit, while the capital market puts (common knowledge) probabilities α that the borrower is a good borrower and $1 - \alpha$ that she is a bad borrower. All other parameters are common knowledge between the borrower and the lenders.

To make things interesting, let us assume that under asymmetric information, the lenders are uncertain about whether the project should be funded:

$$p_H(R - \frac{B_H}{\Delta p}) < I < p_H(R - \frac{B_L}{\Delta p}).$$

Assume that investors cannot break even if the borrower shrinks:

$$p_L R < I.$$

(a) Note that the investor cannot finance only good borrowers. Assume that the entrepreneur receives no reward in the case of failure (this is indeed optimal); consider the effect of rewards R_b in the case of success that are smaller than $B_L/\Delta p$, larger than $B_L/\Delta p$, or between those two values.

(b) Show that there exists $\alpha*$, $0 < \alpha* < 1$, such that

- no financing occurs if $\alpha < \alpha*$
- financing is an equilibrium if $\alpha \geq \alpha*$.

(c) Describe the "cross-subsidies" between types that occur when borrowing is feasible.

7.10 (Extracted from [Tir10], Exercise 6.6). A borrower has asset A and must find financing for a fixed investment $I > A$. As usual, the project yields R (success) or 0 (failure). The borrower is protected by limited liability. The probability of success is p_H or p_L, depending on whether the borrower works or shirks, with $\Delta p \equiv p_H - p_L > 0$. There is no private benefit when working. The private benefit enjoyed by the borrower when shirking is either b (with probability α) or B (with probability $1-\alpha$). At the date of contracting, the borrower knows her private benefit, but the market (which is risk neutral and charges a 0 average rate of interest) does not know it. Assume that $p_L R + B < I$ (the project is always inefficient if the borrower shirks) and that

$$p_H(R - \frac{B}{\Delta p} < I - A < p_H(R - \frac{b}{\Delta p})$$

and

$$[\alpha p_H + (1-\alpha)p_L][R - \frac{b}{\Delta p}] < I - A.$$

(a) Interpret conditions (6.4) and (6.5) and show that there is no lending in equilibrium.
(b) Suppose now that the borrower can at cost $r(x) = rx$ (which is paid from the cash endowment A) purchase a signal with quality $x \in [0, 1]$. (This quality can be interpreted as the reputation or the number of rating agencies that the borrower contracts with.) With probability x, the signal reveals the borrower's type (b or B) perfectly; with probability $1 - x$, the signal reveals nothing. The financial market observes both the quality x of the signal chosen by the borrower and the outcome of the signal (full or no information). The borrower than offers a contract that gives the borrower R_b and the lenders $R - R_b$ in the case of success (so, a contract is the choice of an $R_b \in [0, R]$). The timing is summarized as follows:

1. Borrower chooses quality of signal x (this quality is observed by the capital market).
2. Borrowers type is revealed with probability x. Nothing revealed with probability $1 - x$.
3. Borrower goes to the capital market.

Look for a pure strategy, separating equilibrium, that is, an equilibrium in which the two types pick different signal qualities.

- Argue that the bad borrower (borrower B) does not purchase a signal in a separating equilibrium.
- Argue that the good borrower (borrower b) borrows under the same conditions regardless of the signal's realization, in a separating equilibrium.
- Show that the good borrower chooses signal quality $x \in (0, 1)$ given by

$$A = x(A - rx) + (1-x)\left(p_L(R - \frac{I - A + rx}{p_H}) + B\right).$$

- Show that this separating equilibrium exists only if r is "not too large".

8 Time-Continuous Model

8.1. Let W be a Wiener process. Find the expressions for

$$d(W^2) \text{ and } (dW)^2.$$

8.2. Using Ito's Lemma, compute

$$E[e^{\mu+\sigma W(T)}].$$

8.3. Suppose a stack follows the geometric Brownian motion

$$\frac{dS}{S} = \mu dt + \sigma dW(t).$$

8.4. Assume

$$d\sqrt{V_t} = \beta\sqrt{V_t}dt + \delta dW, \text{ where } \beta, \delta \in \mathbb{R}.$$

Show that

$$dV_t = [\delta^2 - 2\beta V_t]dt + 2\delta\sqrt{V_t}dW.$$

8.5. Assume

$$\frac{dS}{S} = \mu dt + \sigma dW,$$

(a) Show that S^5 follows the geometric Brownian motion.
(b) Find $E[S^5(T)]$.
(c) Find the price of $S^5(T)$ at time t.

T. Hens, M. O. Rieger, *Solutions to* Financial Economics,
Springer Texts in Business and Economics,
https://doi.org/10.1007/978-3-662-59889-4_8

8.6. Show that

$$E(S(t)) \equiv e^{S(t)-\frac{1}{2}\int_0^t S^2(T)\sigma^2}d$$

satisfies

$$dE(S(t)) = E(S(t))dS(t),$$

where

$$\frac{dS}{S}\mu dt + \sigma dW.$$

8.7.

(a) Let $M = \{M(t)\}_{t=0,\ldots,T}$ be a martingale with $M(T) = 0$. Show that $M(t) = 0$ for $t = 0, \ldots, T \ldots$.
(b) In the coin toss space, consider the process M such that

$$M(t) \equiv \sum_{s=1}^{t} R[Z_s] - \alpha\, t \text{ for } t = 1, \ldots, T, \qquad \mathrm{M}(0) = 0.$$

For what values of α is M a martingale?

8.8. Let us consider the stochastic volatility model of 8.8.4 in the textbook, i.e.

$$dS(t) = \mu S(t)\,\mathrm{d}t + \sqrt{\sigma(t)}S(t)\,\mathrm{d}B_1(t),$$
$$d\sigma(t) = \alpha(\sigma(t))\,\mathrm{d}t + \beta(\sigma(t))\,\mathrm{d}B_2(t),$$

where $\alpha(\sigma(t)) = \theta(\omega - \sigma(t))$ (where ω, θ are positive constants), β is given and B_1, B_2 are both Brownian motions that may correlate with each other with correlation $\rho \in [-1, +1]$.

What would happen if we choose $\beta = 0$? And what if β is instead a *positive* constant?

Part II

Solutions

1 Introduction

In this part, we provide the solutions to the exercises from Part I. Sometimes we will be very brief (if we think that this is sufficient to understand the solution), sometimes we will be a bit more wordy (if the solution requires some careful argument). We do not write the solutions in a "paper style", but rather in a way one would explain it to a student on a blackboard. We think in this, more vivid way, computations and arguments can be easier understood, even though this sometimes makes derivations look longer than in the condensed style of a scientific paper: a lengthy computation doesn't get easier if we abbreviate it by leaving the "easy" steps to the poor reader...

Finally, if you find alternative solutions, we very much appreciate that: we never claim our solutions to be the best and in particular not the only one!

T. Hens, M. O. Rieger, *Solutions to* Financial Economics,
Springer Texts in Business and Economics,
https://doi.org/10.1007/978-3-662-59889-4_9

2 Decision Theory

2.1

(a) The expected utility of participating in the lottery is

$$\frac{1}{6}\log(10'000 + 6'000'000) + \frac{5}{6}\log(10'000) = 4.46,$$

while the utility of a sure gain of one million is

$$\log(10'000 + 1'000'000) = 6.00.$$

Therefore, a sure gain of one million would be preferred. We could have concluded this also without computations by noting that for a risk-averse utility function $u(\cdot)$ (e.g., logarithmic utility) the expected value of the lottery is always preferred over the lottery itself. Since the expected value of the given lottery is

$$\frac{1}{6}6'000'000 + \frac{5}{6}0 = 1'000'000,$$

we obtain that

$$u(10'000 + 1'000'000) > \frac{1}{6}u(10'000 + 6'000'000) + \frac{5}{6}u(10'000).$$

(b) If the prize is only 61 €, then the expected utility of the lottery is

$$\frac{1}{6}\log(10'000 + 61) + \frac{5}{6}\log(10'000) = 4.00044.$$

T. Hens, M. O. Rieger, *Solutions to* Financial Economics,
Springer Texts in Business and Economics,
https://doi.org/10.1007/978-3-662-59889-4_10

The lottery's certainty equivalent CE satisfies:

$$\log(10'000 + CE) = 4.00044,$$

or equivalently, $CE = 10.14$ €.

The lottery with a fee of 10 € should be accepted if and only if

$$\frac{1}{6}\log([10'000 - 10] + 61) + \frac{5}{6}\log([10'000 - 10]) = \log([10'000 - 10] + CE')$$
$$\geq \log(10'000).$$

The certainty equivalent CE' of the lottery given wealth $10'000 - 10$ is $10.14 > 10$, therefore, the lottery with the fee of 10 € should be accepted.

This result is a bit surprising, since the expected value of the lottery is just a tiny bit better than the sure alternative. Indeed, one can show, that usual amounts of risk aversion on small stakes are difficult to reconcile with expected utility theory.

2.2 Let A, B and C be three lotteries such that $U(A) \geq U(B) \geq U(C)$, where the utility $U(X)$ of the lottery X is given by

$$U(X) = \int_{\mathbb{R}} u(x)dp(x).$$

If p_{mix} is the probability with which the lottery A is played and $1 - p_{mix}$ is the probability with which the lottery C is played, then the expected utility of this mix can be computed as

$$U(p_{mix}A + (1 - p_{mix})C) = p_{mix}\int_{\mathbb{R}} u(a)dp(a) + (1 - p_{mix})\int_{\mathbb{R}} u(c)dp(c)$$
$$= p_{mix}U(A) + (1 - p_{mix})U(C).$$

Hence, in order for an EUT agent to be indifferent between this mix and the lottery B, p_{mix} has to satisfy the following equation:

$$U(B) = p_{mix}U(A) + (1 - p_{mix})U(C),$$

or equivalently,

$$p_{mix} = \frac{U(B) - U(C)}{U(A) - U(C)},$$

provided that,

$$U(A) \neq U(C).$$

If otherwise $U(A) = U(C)$, the solution for p_{mix} is unidentified, i.e. the equation would be satisfied for any p_{mix}. If $U(A) > U(C)$, since $U(A) \geq U(B) \geq U(C)$, it follows that $p_{mix} \in [0, 1]$. Consequently, the Continuity Axiom is indeed satisfied.

2.3 The efficiency of the measure can be determined through computing the expected utility of playing the lottery of illegal parking under that measure. The measure with the lowest expected utility of illegal parking should be taken.

Therefore, we compute the difference of expected utilities under the two measures:

$$\begin{aligned}
&U(\text{Double Fine}) = p \cdot u(w - 2f) + (1 - p) \cdot u(w), \\
&U(\text{Double Controls}) = 2p \cdot u(w - f) + (1 - 2p) \cdot u(w), \\
&\Delta U = U(\text{Double Fine}) - U(\text{Double Controls}), \\
&= [p \cdot u(w - 2f) + (1 - p) \cdot u(w)] - [2p \cdot u(w - f) + (1 - 2p) \cdot u(w)] \\
&= p[u(w - 2f) - 2u(w - f) + u(w)] \\
&< p[2u(w - f) - 2u(w - f)] = 0,
\end{aligned}$$

where we used that by Jensen's inequality for a strictly concave utility function u, $\frac{1}{2}u(w - 2f) + \frac{1}{2}u(w) < u(w - f)$. Consequently, since

$$U(\text{Double Fince}) < U(\text{Double Controls}),$$

doubling the fee for illegal parking is the more efficient measure.

2.4

(a) The bond is a risk-free asset, hence

$$\mu_{bond} = 0.02, \quad \sigma^2_{bond} = 0.$$

As for the stock,

$$\begin{aligned}
\mu_{stock} &= \frac{1}{2}0.08 + \frac{1}{2}(-0.02) = 0.03, \\
\sigma^2_{stock} &= \frac{1}{2}(0.08 - 0.03)^2 + \frac{1}{2}(-0.02 - 0.03)^2 = 0.0025.
\end{aligned}$$

(b) An investor with the mean-variance utility $U(\mu, \sigma^2) = \mu - \alpha\sigma^2$ would value these two assets as follows:

$$U(bond) = 0.02, \quad U(stock) = 0.03 - \alpha \cdot 0.0025.$$

Therefore, to have him being indifferent between the bond and the stock, the parameter α should be satisfy the condition

$$0.03 - \alpha \cdot 0.0025 = 0.02,$$

or equivalently, $\alpha = 4$.

(c) If an investor spends a share λ of wealth on buying the stock and $(1 - \lambda)$ on buying the bond, then his portfolio delivers a return of

$$(1 - \lambda) \cdot 0.02 + \lambda \cdot 0.08 = 0.02 + 0.06 \cdot \lambda$$

in boom, and

$$(1 - \lambda) \cdot 0.02 + \lambda \cdot (-0.02) = 0.02 - 0.04 \cdot \lambda.$$

in recession. For the mean and the variance of the portfolio we obtain:

$$\mu_\lambda = \frac{1}{2}(0.02 + 0.06 \cdot \lambda) + \frac{1}{2}(0.02 - 0.04 \cdot \lambda) = 0.02 + 0.01 \cdot \lambda,$$
$$\sigma_\lambda^2 = \lambda^2 \cdot \sigma_{stock}^2 = \lambda^2 \cdot 0.0025.$$

The utility of the investor as function of λ is therefore:

$$U(\lambda) = 0.02 + 0.01 \cdot \lambda - 0.0025\alpha\lambda^2.$$

Since $U(\lambda)$ is strictly concave with respect to λ, the optimal λ^{opt} is a unique solution of the FOC:

$$\frac{\partial U(\lambda)}{\partial \lambda} = 0.01 - 0.005\alpha\lambda,$$

or equivalently, $\lambda^{opt} = \frac{2}{\alpha}$.

2.5 Prospect Theory predicts risk-seeking behaviour in losses, hence Daniel Kahneman would put both credit cards into one wallet in order to reduce the probability of the loss. On the other hand, expected utility theory with a concave utility function predicts diversification of risk, therefore, Daniel Bernoulli would hold one credit card in each wallet.

Indeed, the two possible choices, namely two wallets each having a card or one wallet having both cards and one empty wallet, are equivalent to two lotteries. Denote by lottery A the first choice and by lottery B the second choice. Let p be the probability of loosing a wallet. The loss of one wallet is independent of the loss of the other wallet, i.e. we assume pickpocket thieves (typically stealing just one wallet) and not robbers who would steal both wallets at once. Then the lottery A is given as

Probability	p^2	$2p(1-p)$	$(1-p)^2$
Outcome	$-2L$	$-L$	0

and the lottery B is given as

Probability	p	$1-p$
Outcome	$-2L$	0

These two lotteries translate into the following utilities:

$$U(A) = v(-2L) \cdot p^2 + v(-L) \cdot 2p(1-p) + v(0) \cdot (1-p)^2,$$
$$U(B) = v(-2L) \cdot p + v(0) \cdot (1-p)$$

with

$$U(A) - U(B) = p(1-p)[-v(-2L) + 2v(-L) - v(0)],$$

where $v(\cdot)$ is the value function. The value function is concave on the entire domain for Bernoulli and is convex over losses for Kahneman. Therefore, $U(A) > U(B)$ for Daniel Bernoulli and $U(A) < U(B)$ for Daniel Kahneman (mathematically, this problem is very similar to 2.3.).

2.6 Yes, it can. The standard form of Prospect Theory assumes that for a lottery A with n outcomes $x_1, \ldots, x_n$ and probabilities $p_1, \ldots, p_n$:

$$PT(A) = \sum_{i=1}^{n} v(x_i),$$

where the value function is given by

$$v(x) = \begin{cases} x^\alpha, & x \geq 0, \\ -\lambda(-x)^\beta, & x < 0 \end{cases}$$

and the probability weighting function is given by

$$w(p) = \frac{p^{\gamma}}{(p^{\gamma} + (1-p)^{\gamma})^{1/\gamma}}.$$

Taking, e.g., $\lambda = 2.25$, $\alpha = \beta = 0.8$ and $\gamma = 0.7$, it can be computed that

$$\begin{aligned} PT(lottery) &= w\left(\frac{1}{1'000'001}\right) v(1'000'000-2) + w\left(\frac{1'000'000}{1'000'001}\right) v(-2) \\ &= 0.064 > 0. \end{aligned}$$

2.7 Taking, e.g., $\lambda = 2.25$, $\alpha = \beta = 0.8$ and $\gamma = 0.7$, we compute the CPT values of the four lotteries:

$$\begin{aligned} &CPT(A) \\ &= w(0.01)v(0) + (w(0.67) - w(0.01))v(2400) + (1 - w(0.67))v(2500) \\ &= 493.98, \\ &CPT(B) = v(2400) = 506.02, \\ &CPT(C) = w(0.67)v(0) + (1 - w(0.67))v(2500) = 224.01, \\ &CPT(D) = w(0.66)v(0) + (1 - w(0.66))v(2400) = 220.44. \end{aligned}$$

It follows that $A \prec B$ and $D \prec C$, which is the preference pattern that we wanted to explain with the theory.

2.8 Yes, this can happen in PT. Consider a lottery with an outcome 1 with 10% and 2 with 90% probability. Assume that $v(x) = x$ and take the probability weighting function $w(p) = p^{1/10}$. Then the PT value of this lottery is

$$PT(X) = 0.1^{\frac{1}{10}} + 0.9^{\frac{1}{10}} \cdot 2 \approx 0.79 + 1.96 = 1.96,$$

and the certainty equivalent is

$$CE(X) = v^{-1}(PT(X)) = 2.75.$$

In this example the certainty equivalent of the lottery (2.75) is higher than its highest outcome (2). This comes from the fact that the probability of the unlikely event is so much overweighted that it dominates the result.

The violation of internality cannot happen in the case of the cumulative prospect theory and the normalized prospect theory:

$$CPT(X) = \sum_{i=1}^{I}(w(F_i) - w(F_{i-1}))v(x_i) =: \sum_{i=1}^{I} w_i^{CPT} v(x_i),$$

$$NPT(X) = \sum_{i=1}^{I} \frac{w(p_i)}{\sum_{j=1}^{I} w(p_j)} v(x_i) =: \sum_{i=1}^{I} w_i^{NPT} v(x_i).$$

It is straightforward to check that $\sum_{i=1}^{I} w_i^{CPT(NPT)} = 1$ in both cases. Consequently, both in CPT and NPT the utility of a lottery cannot become bigger than the utility of its largest outcome:

$$\begin{aligned}\sum_{i=1}^{I} w_i^{CPT(NPT)} v(x_i) &\leq \max_i v(x_i) \sum_{i=1}^{I} w_u^{CPT(NPT)} \\ &= \max_i v(x_i) \cdot 1 \\ &\leq v(\max_i x_i).\end{aligned}$$

Therefore, due to strict monotonicity of v, also the certainty equivalent of a lottery,

$$CE^{CPT}(X) = v^{-1}(CPT(X)),$$
$$CE^{NPT}(X) = v^{-1}(NPT(X)),$$

cannot become bigger than the largest outcome.

2.9 For the given set of parameters the PT value of the described lottery is

$$PT(X) = w(0.5)v(100) + w(0.5)w(-100) = 0.5 \cdot 100^{0.8} - 0.5 \cdot 2.25 \cdot \lambda \cdot 100.$$

Therefore, the values of λ for which the person is not willing to participate in this lottery are such that

$$0.5 \cdot 100^{0.8} - 0.5 \cdot 2.25 \cdot \lambda \cdot 100 < 0,$$

or equivalently, $0.177 < \lambda$.

More generally, for any values of α and β, such that $\alpha < \beta < 1$, the PT value of the lottery is

$$PT(X) = w(0.5)v(100) + w(0.5)w(-100) = 0.5 \cdot 100^{\alpha} - 0.5 \cdot 2.25 \cdot \lambda \cdot 100^{\beta}.$$

The person is then loss averse for any λ that satisfies

$$\frac{100^{\alpha-\beta}}{2.25} < \lambda.$$

Since for $\alpha < \beta$, $\frac{100^{\alpha-\beta}}{2.25} < \frac{1}{2.25} < 1$, there exists some $\lambda < 1$ such that the person is loss averse for that value of λ.

2.10 The CPT value of the lottery can be computed as follows:

$$\begin{aligned} CPT(p) &= \int_{\mathbb{R}} v(x)dw(F(x)) \\ &= \int_{\mathbb{R}} v(x)w'(F(x))F'(x)dx \\ &= \int_{\mathbb{R}} \frac{x}{2\sqrt{F(x)}} p(x)dx, \end{aligned}$$

where the cumulative probability is

$$F(x) = \int_{-\infty}^{x} p(t)dt = \begin{cases} 0, & \text{if } x < 0, \\ \frac{x^2}{2}, & \text{if } 0 \le x < 1, \\ -\frac{x^2}{2} + 2x - 1, & \text{if } 1 \le x < 2, \\ 1, & \text{if } 2 \le x. \end{cases}$$

Hence,

$$\begin{aligned} CPT(p) &= \int_0^1 \frac{x}{\sqrt{2}x} x dx + \int_1^2 \frac{x}{2\sqrt{-\frac{x^2}{2} + 2x - 1}} (2-x)dx \\ &= \frac{1}{2\sqrt{2}} + \int_1^2 x d\sqrt{-\frac{x^2}{2} + 2x - 1} \\ &= \frac{1}{2\sqrt{2}} + x\sqrt{-\frac{x^2}{2} + 2x - 1}\Bigg|_1^2 - \int_1^2 \sqrt{-\frac{x^2}{2} + 2x - 1dx} \\ &= -\frac{1}{2\sqrt{2}} + 2 - \int_{-1}^{0} \frac{1}{\sqrt{2}} \sqrt{2 - y^2 dy} \end{aligned}$$

$$= -\frac{1}{2\sqrt{2}} + 2 - \frac{1}{2\sqrt{2}} \left(y\sqrt{2-y^2} + 2\arcsin\frac{y}{\sqrt{2}} \right) \Big|_{-1}^{0}$$

$$= -\frac{1}{2\sqrt{2}} + 2 - \frac{1}{2\sqrt{2}} \left(1 + 2 \cdot \frac{\pi}{4} \right) \approx 0.74$$

with $y = x - 2$. The lottery's certainty equivalent CE is the solution of

$$v(CE) = CPT(p).$$

Hence, given that $v(CE) = CE$, we obtain $CE = CPT(p) = 0.74$. As for the expected value of the lottery,

$$EV(p) = \int_{\mathbb{R}} v(x)p(x)dx = \int_0^1 x^2 dx + \int_1^2 x(2-x)dx = 1.$$

We see that the certainty equivalent is smaller than the expected value, hence the person behaves risk-averse. This is due to the weighting function $w(F) = \sqrt{F}$, which overweights events with small probability. Without weighing (when $w(F) = F$), the value function $v(x) = x$ alone would indicate risk neutrality, i.e., the expected utility would be equal to the expected value.

2.11 Assume δ is the discount factor per period, which we choose to be 1 week. For Jerome we know:

(1) $U(w_0 + 100) > \beta\delta U(w_1 + 110)$

(2) $U(w_0 + 100) < \beta\delta^{16} U(w_1 + 200)$ (4 months $\approx$ 16 weeks)

Classical model $\beta = 1,\ w_0 = w_1 = 0,\ U(w) = w$ gives

$$(1)\ \delta < \frac{U(w_0 + 100)}{U(w_1 + 110)} \implies \delta < \frac{100}{110} = 0.91$$

$$(2)\ \delta^{16} > \frac{U(w_0 + 100)}{U(w_1 + 200)} \implies \delta > \left(\frac{100}{200}\right)^{\frac{1}{16}} = 0.96$$

This is a contradiction! Further, by solving the following for beta

$$(1)\ \frac{10}{11}\frac{1}{\beta} > \delta,$$

$$(2)\ \left(\frac{1}{2}\frac{1}{\beta}\right)^{\frac{1}{16}} < \delta$$

$$\implies \left(\frac{1}{2}\frac{1}{\beta}\right)^{\frac{1}{16}} < \frac{10}{11}\frac{1}{\beta},$$

we can show that $\beta < 0.95$. Thus, Jerome's case cannot be explained by exponential discount, but by hyperbolic discounting.

For Angelika we again start with the same assumptions as above, expect that we use a yearly discounting for δ and we write:

$$(1)\ U(w_0 + 100) < \beta\delta U(w_1 + 120),$$

$$(2)\ U(w_0 + 100) > \beta\delta^{10} U(w_1 + 1000),$$

Assuming linear utility, we again find a contradiction:

$$(1)\ \frac{10}{12} < \delta,$$

$$(2)\ \left(\frac{1}{10}\right)^{\frac{1}{10}} > \delta.$$

Now we will try to explain the preferences with hyperbolic discounting (with $\beta < 1$):

$$(1)\ \frac{10}{12}\frac{1}{\beta} < \delta$$

$$(2)\ \left(\frac{1}{10}\frac{1}{\beta}\right)^{\frac{1}{10}} > \delta \implies \beta > 1.05$$

Since β is larger than 1, we can explain Angelika's behavior neither by exponential nor by hyperbolic discounting. We now assume an utility of the form that was given in the hint, i.e. $U(w) = \min(w, c)$. As an example we set $c = 300$, $w_0 = w_1 = 100$, $w_{10} = 500$. In the following we leave out the utility of the periods where nothing happened, since it has no impact on the results. Now we have the second inequality:

$$(2)\ U(100 + 100) + \delta^{10} U(500) > U(100 + \delta^{10} U(500 + 1000)$$

If we plug in our assumptions, we can see that this inequality holds for any values of δ:

$$\min(200, 300) + \delta^{10} \min(500, 300) > \min(100, 300) + \delta^{10} \min(1500, 300),$$

$$200 + \delta^{10} 300 > 100 + \delta^{10} 300.$$

This comes from the fact that under the assumed utility, the rising wealth level $w_{10} >> w_0$ decreases the marginal utility of more wealth (in this case to zero).

The first inequality is:

$$U(100+100)+\delta U(100) < U(100)+\delta U(100+120),$$
$$\min(200,300)+\delta\min(100,300) < \min(100,300)+\delta\min(220,300,)$$
$$\delta > \frac{5}{6}.$$

Taking into account a growing wealth level can explain Angelikas behavior, assuming the discount factor δ is larger than $\frac{5}{6}$, but neither hyperbolic nor classical time discounting alone are sufficient.

2.12 To construct such a lottery L, it is enough to use a binary lottery with 50% chance each for winning an amount x or loosing y. Let us say that your initial wealth is w. Then the utility of playing the lottery once is

$$U(L) = \frac{1}{2}u(w+x) + \frac{1}{2}u(w-y).$$

Playing the lottery twice is equivalent to a lottery that gives $w + 2x$ with probability 25%, $w+x-y$ with probability 50% and $w-2y$ with probability 25%:

$$U(2L) = \frac{1}{4}u(w+2x) + \frac{1}{2}u(w+x-y) + \frac{1}{4}u(w-2y).$$

Setting, e.g., $u(w) = w^\alpha$, $\alpha = 0.7$, $x = 0.03 \cdot w$ and $y = 0.04 \cdot w$ leads to $U(L) = 0.996 \cdot w^\alpha < u(w)$ and $U(2L) = 1.229 \cdot w^\alpha > u(w)$ as required.

Now, what about the paradox? In Samuelson's argument he assumes that the risk attitude of the person is the same, even after playing the lottery several times. Playing the lottery several times, however, most likely would result in a subsequent change of the wealth level. Therefore, the risk attitude would still (approximately) be the same only if either the amount on stake is negligibly small, or if the utility function has very special features, namely linear or CARA. In Samuelson's original paper (1964) the amounts are negligible small, but in the application to investment decisions, this is not the case. This resolves the "paradox".

2.13

(a) For the utility of asset 1 we have:

$$V_{KT}(R^1) = \frac{1}{2}0.05^{\alpha^+} + \frac{1}{2}0.1^{\alpha^+} = 0.102.$$

Hence, its certainty equivalent CE_1 is given by:

$$0.102 = CE_1^{\alpha^+},$$

or equivalently,

$$CE_1 = (0.184)^{\frac{1}{\alpha^+}} = 0.0745.$$

In the same way the other certainty equivalents are obtained: $CE_2 = -0.0117$ and $CE_3 = -0.0745$.

(b) To find the parameters of the quadratic function the following relation is used:

$$V_{DHM}(R^i) = V_{DHM}(CE_i)$$

with $i = 1, 2, 3$.

For $i = 1$ we have:

$$\frac{1}{2}\left(0.05 - \frac{\alpha^+}{2}0.05^2\right) + \frac{1}{2}\left(0.1 - \frac{\alpha^+}{2}0.1^2\right) = CE_1 + \frac{\alpha^+}{2}CE_1^2.$$

Therefore, $\alpha^+ = 1.46$. In the same way, from $i = 2$ and $i = 3$, $\alpha^- = -1.46$ and $\beta = 1.92$ is obtained.

2.14

(a) The optimal allocation of mean-variance maxmizer into a single risky asset is:

$$\lambda^* = \frac{\mu - R_f}{\gamma\sigma^2}$$

The mean and the variance are

$$\mu = 0.0275, \text{VAR} = 0.00541$$

Therefore

$$\lambda^* = 1.61, \lambda_0^* = -0.61$$

(b) R_s is the return of the stock in state s. R_λ is the return where the investor invests a fraction λ into the risky asset and $1 - \lambda$ in the risk-free asset.

The value function is:

$$v(R_\lambda) = \begin{cases} (R_\lambda - RP) - \frac{\alpha^+}{2}(R_\lambda - RP)^2 & \text{if } R_\lambda > RP \\ \beta\left((R_\lambda - RP) - \frac{\alpha^-}{2}(R_\lambda - RP)^2\right) & \text{if } R_\lambda \leq RP \end{cases}$$

Since $R_\lambda = (1-\lambda)R_f + \lambda R$ and $RP = R_f$, it follows $R_\lambda - RP = \lambda(R - R_f)$. Therefore

$$v(R_\lambda) = \begin{cases} (\lambda(R - R_f)) - \frac{\alpha^+}{2}(\lambda(R - R_f))^2 & \text{if } \lambda R_\lambda > \lambda R_f \\ \beta\left((\lambda(R - R_f)) - \frac{\alpha^-}{2}(\lambda(R - R_f))^2\right) & \text{if } \lambda R_\lambda < \lambda R_f \end{cases}$$

The first order condition of the prospect utility maximization problem with 2 states

$$w(p_1)\frac{\partial v(R_1)}{\partial \lambda} = -w(p_2)\frac{\partial v(R_1)}{\partial \lambda},$$

where $w(p_s)$ is the probability weighting. Assuming $\lambda R_1 > \lambda R_f > \lambda R_2$ (this is given if $\lambda > 0$) and plugging v into the first order condition we get

$$w(p_1)\left[R_1 - R_f - \alpha^+\lambda\left(R_1 - R_f\right)^2\right] = -w(p_2)\beta\left[R_2 - R_f - \alpha^-\lambda\left(R_2 - R_f\right)^2\right].$$

Solving for λ leads to

$$\lambda^* = \frac{w(p_1)(R_1 - R_f) + w(p_2)\beta(R_2 - R_f)}{w(p_1)\alpha^+(R_1 - R_f)^2 + w(p_2)\beta\alpha^-(R_2 - R_f)^2}.$$

To ensure that we have indeed a maximum, the second order condition must be satisfied, i.e.

$$\frac{\partial^2 V}{\partial \lambda^2} = w(p_1)(-1)\alpha^+(R_1 - R_f)^2 + w(p_2)(-1)\beta\alpha^-(R_2 - R_f)^2 < 0.$$

For our example we obtain

$$\lambda^* = 1.99, \qquad \frac{\partial^2 V}{\partial \lambda^2} = 0.005 > 0.$$

In other words we have found a minimum, but we should have maximized the utility! Since the second derivative of V is positive, V is strictly convex. Because of that and the fact that the global minimum of V is at 1.99 it follows under the short selling constraints that the optimal decision of the investor is to invest everything in the risk-free asset (i.e. $\lambda^* = 0$ and $\lambda_0^* = 1$).

(c) This problem can be solved analogously. The results are

$$\lambda^* = 0.364, \qquad \lambda_0^* = 0.636, \qquad \frac{\partial^2 V}{\partial \lambda^2} = -0.00824 < 0.$$

In this case we have indeed a maximum.

2.15

(a) Assume that w is your yearly income. Consider the game L which pays you either $(1 - x)w$ or $2w$ with equal probabilities 50%. You need to choose x such that you are indifferent between taking part in the game L and having a secure payment w.

The answer depends solely on individual's preferences.

(b) For a given utility function $u(c)$ a **coefficient of relative risk aversion** is defined as

$$R(c) = -c\frac{u''(c)}{u'(c)}.$$

CRRA is a class of utility functions with a constant relative risk aversion. It can be shown that each CRRA utility may be represented as

$$u^{\alpha}(c) = \frac{c^{1-\alpha}}{1-\alpha},$$

where $\alpha \neq 1$ is the constant coefficient of relative risk aversion plus the function $u'(c) = \ln c$.

To find your CRRA, you need to find α, such that for the utility $u^{\alpha}(\cdot)$ you are indifferent between w and L:

$$\frac{w^{1-\alpha}}{1-\alpha} = 0.5\frac{((1-x)w)^{1-\alpha}}{1-\alpha} + 0.5\frac{(2w)^{1-\alpha}}{1-\alpha},$$

or equivalently,

$$2 = (1-x)^{1-\alpha} + 2^{1-\alpha}.$$

α can be found numerically. For example, for $x = 25\%$ it is $\alpha = 2.91$.

(c) For the mean-variance investor the problem has to be rewritten in terms of net returns. We get a new game L', which pays you either $(-x)$ or 1 of net return with equal probabilities 50%. The expected value and the variance of L' are

$$\mu(L') = 0.5 \cdot (-x) + 0.5 \cdot 1 = 0.5 - 0.5x,$$

$$\sigma^2(L') = \mathbb{E}(L' - \mu(L'))^2 = 0.25(1+x)^2,$$

correspondingly. The sure alternative has an expected value and a standard deviation of 0.

A mean-variance investor must be indifferent between L' and the secure alternative:

$$U(L') = U(0) \Longleftrightarrow \mu(L') - \frac{\gamma}{2}\sigma^2(L') = 0 \Longleftrightarrow \gamma = 2\frac{\mu(L')}{\sigma^2(L')} = 4\frac{1-x}{(1+x)^2}.$$

For $x = 25\%$, we get risk-aversion $\gamma = 1.92$.

(d) Like a mean-variance investor, a prospect theory maximizer deals with returns. Therefore, the prospect theory investor must be indifferent between L' and the secure alternative:

$$v(L') = v(0) \iff 0 = \frac{1}{2}(-\beta)(0+x)^{\alpha^-} + \frac{1}{2}(1-0)^{\alpha^+}$$

$$\iff x = \beta^{-1/\alpha^-} = 0.4.$$

(e) Denote by r_s the return on stocks, σ_s the standard deviation of stocks and r_0 the return of bonds. Because bonds are risk-free, their standard deviation equals zero. If the investor invests a fraction λ of his wealth into stocks and $1-\lambda$ into bonds, then the return of his portfolio is $\lambda r_s + (1-\lambda) r_0$, while the standard deviation is $\lambda \sigma_s$. Thus, the maximization problem of the mean-variance investor becomes

$$\max_{\lambda} \lambda r_s + (1-\lambda) r_0 - \frac{\gamma}{2}\lambda^2 \sigma_s^2.$$

The FOC of this problem is

$$(r_s - r_0) - \gamma \lambda^* \sigma_s^2 = 0,$$

therefore,

$$\gamma = \frac{r_s - r_0}{\lambda^* \sigma_s^2},$$

where λ^* is the solution of optimization problem. Plugging the values above, we obtain $\gamma = 2.9$.

(f) The CRRA investor must be indifferent between the risky lottery and the secure alternative. In the case, when the background wealth is added, it means that

$$U(w + 0.5w) = U(L + 0.5w) \iff$$

$$\frac{(w+0.5w)^{1-\alpha}}{1-\alpha} = 0.5\frac{((1-x)w + 0.5w)^{1-\alpha}}{1-\alpha} + 0.5\frac{(2w+0.5w)^{1-\alpha}}{1-\alpha},$$

$$\iff 2 \cdot 1.5^{1-\alpha} = (1.5-x)^{1-\alpha} + 2.5^{1-\alpha}.$$

For $x = 25\%$, we get $\alpha = 4.25$.

3 Two-Period Model: Mean-Variance Approach

3.1

(a) The minimum–variance portfolio is the portfolio of risky assets with the minimal portfolio variance:

$$(\lambda_1^{MV}, \lambda_2^{MV}) = \arg\min_{\lambda} \sigma_\lambda^2, \qquad \text{s.t. } 0 \le \lambda_1,\ 0 \le \lambda_2,\ \lambda_1 + \lambda_2 = 1, \tag{3.1}$$

where the variance of portfolio λ is

$$\sigma_\lambda^2 = \sigma^2(\lambda_1 R_1 + \lambda_2 R_2) = \lambda^T COV \lambda.$$

Since $\lambda_2^{MV} = 1 - \lambda_1^{MV}$, the dimension of the problem (3.1) can be reduced to one:

$$\lambda_1^{MV} = \arg\min_{\lambda_1} (\lambda_1, 1 - \lambda_1) COV \begin{pmatrix} \lambda_1 \\ 1 - \lambda_1 \end{pmatrix}, \qquad \text{s.t. } 0 \le \lambda_1 \le 1. \tag{3.2}$$

For values of COV as given above, the optimization problem (3.2) reads as follows:

$$\lambda_1^{MV} = \arg \min_{0 \le \lambda_1 \le 1} \left(\lambda_1\ 1 - \lambda_1\right) \begin{pmatrix} 2 & -1 \\ -1 & 4 \end{pmatrix} \begin{pmatrix} \lambda_1 \\ 1 - \lambda_1 \end{pmatrix}.$$

From the FOC for unconstrained minimization problem in (3.2),

$$4\lambda_1^{MV} - 2(1 - \lambda_1^{MV}) + 2\lambda_1^{MV} - 8(1 - \lambda_1^{MV}) = 0,$$

T. Hens, M. O. Rieger, *Solutions to* Financial Economics,
Springer Texts in Business and Economics,
https://doi.org/10.1007/978-3-662-59889-4_11

we find that $\lambda_1^{MV} = 0.625$ and $\lambda_2^{MV} = 1 - \lambda_1^{MV} = 0.375$. Since these values meet $0 \leq \lambda_1^{MV} \leq 1$ and $0 \leq \lambda_2^{MV} \leq 1$, short selling constraints in (3.2) are satisfied, hence $(\lambda_1^{MV}, \lambda_2^{MV})$ is the solution of the original problem (3.1), or equivalently, the minimum-variance portfolio.

The tangent portfolio is the portfolio of risky assets with the maximum Sharpe ratio, i.e. providing maximum excess return for a unit of risk:

$$\lambda^{Tg} = \arg\max_{\lambda} \frac{\mu_\lambda - R_f}{\sigma_\lambda}, \quad \text{s.t. } 0 \leq \lambda_1,\ 0 \leq \lambda_2,\ \lambda_1 + \lambda_2 = 1. \tag{3.3}$$

The easiest way to find the tangent portfolio is to maximize the investor i's utility and use the result of the Two-Fund Separation Theorem, i.e., that the optimal portfolio of investor satisfies $\lambda^{i,opt} = (\lambda_0^{i_o pt}, (1 - \lambda_0^{i_o pt})\lambda^{Tg})$. The optimal investment strategy of agent i can be found as follows:

$$\lambda^{i,opt} = \arg \max_{\lambda \in \mathbb{R}^{2+1}} U^i(\mu_\lambda, \sigma_\lambda^2) = \mu_\lambda - \frac{\rho^i}{2}\sigma_\lambda^2, \quad \text{s.t} \sum_{k=0}^{2} \lambda_k = 1, \tag{3.4}$$

where λ_0 is the fraction of wealth invested into the riskless asset. From the budget constraint, $\lambda_0 = 1 - \sum_{k=1}^{2} \lambda_k$, hence the constrained problem (3.4) is equivalent to:

$$(\lambda_1^{i,opt}, \lambda_2^{i,opt}) = \arg\max_{\lambda \in \mathbb{R}^2} (\mu - R_f \mathbb{I})^T \lambda - \frac{\rho^i}{2} \lambda^T COV \lambda.$$

Writing FOCs, we find that

$$COV \begin{pmatrix} \lambda_1^{i,opt} \\ \lambda_2^{i,opt} \end{pmatrix} = \frac{1}{\rho^i}(\mu - R_f \mathbb{I}), \tag{3.5}$$

or equivalently,

$$\begin{pmatrix} \lambda_1^{i,opt} \\ \lambda_2^{i,opt} \end{pmatrix} = COV^{-1} \frac{1}{\rho^i}(\mu - R_f \mathbb{I}).$$

The normalized optimal portfolio is then indifferent of i and coincides with the Tangent Portfolio:

$$\lambda_k^{Tg} = \frac{\lambda_k^{i,opt}}{\sum_{l=1}^{2} \lambda_l^{i,opt}} \quad k = 1, 2.$$

For the parameters of the exercise, the first-order conditions (3.5) read as follows:

$$\begin{pmatrix} 2 & -1 \\ -1 & 4 \end{pmatrix} \begin{pmatrix} \lambda_1^{i,opt} \\ \lambda_2^{i,opt} \end{pmatrix} = \frac{1}{\rho^i} \begin{pmatrix} 3 \\ 5.5 \end{pmatrix}.$$

Solving it with respect to $\lambda_1^{i,opt}$ and $\lambda_2^{i,opt}$, we obtain that $\lambda_1^{i,opt} = 5/2\rho^i$ and $\lambda_2^{i,opt} = 2/\rho^i$, both are non-negative. The tangent portfolio is then

$$\lambda_1^{Tg} = \frac{\lambda_1^{i,opt}}{\lambda_1^{i,opt} + \lambda_2^{i,opt}} = \frac{5}{9} = 0.556, \quad \lambda_2^{Tg} = \frac{\lambda_2^{i,opt}}{\lambda_1^{i,opt} + \lambda_2^{i,opt}} = \frac{4}{9} = 0.444.$$

(b) As discussed above, when choosing his investment portfolio, investor i solves the utility maximization problem (3.4). The implied returns can be attained by solving the FOCs (3.5) with respect to μ at given portfolio weights, covariance matrix and riskfree rate:

$$\begin{aligned} \mu^{i,implied} &= R_f \mathbb{I} + \rho^i COV \lambda^i \\ &= \begin{pmatrix} 2\% \\ 2\% \end{pmatrix} + \begin{pmatrix} 2 & -1 \\ -1 & 4 \end{pmatrix} \begin{pmatrix} 0.5\% \\ 0.3\% \end{pmatrix} \\ &= \begin{pmatrix} 2\% \\ 2\% \end{pmatrix} + \begin{pmatrix} 0.7\% \\ 0.7\% \end{pmatrix} = \begin{pmatrix} 2.7\% \\ 2.7\% \end{pmatrix}. \end{aligned}$$

(c) According to the CAPM,

$$\mu_k - R_f = \beta_{k,M}(\mu^M - R_f), \tag{3.6}$$

where

$$\beta_{k,M} = \frac{cov(R_k, R^M)}{\sigma_M^2}, \quad k = 1, 2.$$

To find the beta-factors of the risky assets, we have to calculate covariances of the risky asset returns with the market portfolio:

$$\begin{aligned} \sigma_{1,M}^2 &= cov(R_1, R^M) \\ &= cov(R_1, \lambda_1^M R_1 + \lambda_2^M R_2) \\ &= \lambda_1^M \sigma_1^2 + \lambda_2^M \sigma_{1,2} \\ &= 0.4 \cdot 2\% + 0.6 \cdot (-1\%) = 0.2\%, \end{aligned}$$

respectively,

$$\begin{aligned}\sigma_{2,M}^2 &= COV(R_2, R^M)\\ &= cov(R_2, \lambda_1^M R_1 + \lambda_2^M R_2)\\ &= \lambda_2^M \sigma_2^2 + \lambda_1^M \sigma_{1,2}\\ &= 0.6 \cdot 4\% + 0.4 \cdot (-1\%) = 2\%,\end{aligned}$$

as well as the variance of the market portfolio:

$$\begin{aligned}\sigma_M^2 = \sigma^2(\lambda_1^M R_1 + \lambda_2^M R_2) &= (\lambda_1^M)^2\sigma_1^2 + (\lambda_2^M)^2\sigma_2^2 + 2\lambda_1^M\lambda_2^M\sigma_{1,2}^2\\ &= 0.4^2 \cdot 2\% + 0.6^2 \cdot 4\% + 2 \cdot 0.4 \cdot 0.6 \cdot (-1)\%\\ &= 1,28\%.\end{aligned}$$

Thus, the beta factors of the risky assets are:

$$\beta_{1,M} = \frac{\sigma_{1,M}^2}{\sigma_M^2} = \frac{0.2}{1.28} = 0.1562$$

and

$$\beta_{2,M} = \frac{\sigma_{2,M}^2}{\sigma_M^2} = \frac{2}{1.28} = 1.5625.$$

The expected returns of the risky assets can be then found from equality (3.6):

$$\mu_1 = R_f + \beta_{1,M}(\mu^M - R_f) = 2\% + 0.1562 \cdot 3\% = 2.47\%,$$

respectively

$$\mu_2 = R_f + \beta_{2,M}(\mu^M - R_f) = 2\% + 1.5625 \cdot 3\% = 6.69\%.$$

3.2

(a) The data for the indices is given as price levels. Thus, we first have to calculate the corresponding returns. We can then plot the risk-return tradeoff. Using the MATLAB software:

```
% Cleaning and data import
clear all;
% You might have to specify the current directory
  to make the excel-read
```

```
% possible
% cd('C:\Users\Muster\Desktop\ex3.m')

[data,text]=xlsread('ex_3_extra1.xls');
Rf=data(2:end,13);
indices=data(:,1:12);
IndReturn=(indices(2:end,:)-indices(1:end-1,:))./
 indices(1:end-1,:);
IndNames=text(1,2:13);

%% a)
Means = mean(IndReturn)';
Sigmas = std(IndReturn)';
plot(Sigmas,Means,'o');
xlabel('Standard Deviation');
ylabel('Mean');
h = lsline; % Add a regression line
set(h(1),'color','r')
saveas(gcf,'TradeoffAll.pdf');
% To calculate geometric mean
GeoMeans = geomean(IndReturn+1)-1;
% To calculate annual returns e.g. using arithmetic means
AnnualRet = (Means+1).^12-1;
% To show results a bit nicer use
disp(['Index' IndNames ; 'Mean' num2cell(Means)' ; 'SD'
 num2cell(Sigmas)';
    'Ann. Mean' num2cell(AnnualRet)'])  )
```

The results are shown in Fig. 3.1. We can see that for most of the included country indices higher risk seems to be associated with higher returns. To see this, we have added a linear regression of the points given to the plot. Nevertheless, the points are spread quite wide and we observe one country with very low returns compared to its standard deviation (Japan). The calculated means and standard deviations are shown in Table 3.1.

(b)
```
%% b)
figure(2)
for i = 1:12
    % Plot histogram
    subplot(4,3,i);
    histogram(IndReturn(:,i));
    title(IndNames(1,i));
    axis([-inf inf 0 inf ]);
end
saveas(figure(2),'Histograms.pdf');
```

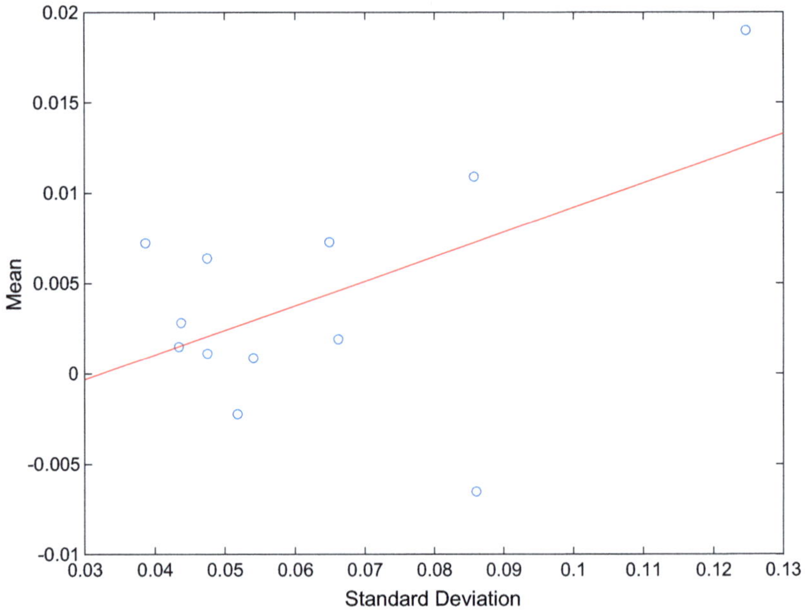

Fig. 3.1 Mean and standard-deviation of given indices

Table 3.1 Mean and standard-deviation of returns

Index	Australia	Canada	China	Germany	Greece	Hong Kong
Mean	0.7185%	0.6343%	1.0873%	0.1886%	−0.6547%	0.7238%
SD	3.8711%	4.7596%	8.5747%	6.6225%	8.6121%	6.4958%
Ann. Mean	8.9706%	7.8823%	13.8568%	2.2874%	−7.5797%	9.0395%
Index	Japan	Switzerl.	Turkey	UK	USA	France
Mean	−0.2274%	0.1457%	1.8962%	0.2787%	0.1087%	0.0828%
SD	5.1869%	4.3510%	12.4733%	4.3813%	4.7639%	5.4106%
Ann. Mean	−2.6952%	1.7623%	25.2844%	3.3959%	1.3118%	0.9984%

```
% Display skewness and kurtosis
display([{'Country'} , {'Skewness'} , {'Kurtosis'} ;
    IndNames' , num2cell(skewness(IndReturn)') ,
    num2cell(kurtosis(IndReturn)')]);
```

Figure 3.2 shows the histograms of the returns. We can see that most of the returns are negatively skewed. To further investigate the distributions, we calculate skewness and kurtosis. All of the countries have kurtosis higher than 3, i.e. the kurtosis of the normal distribution and thus exhibit fat tails. Furthermore,

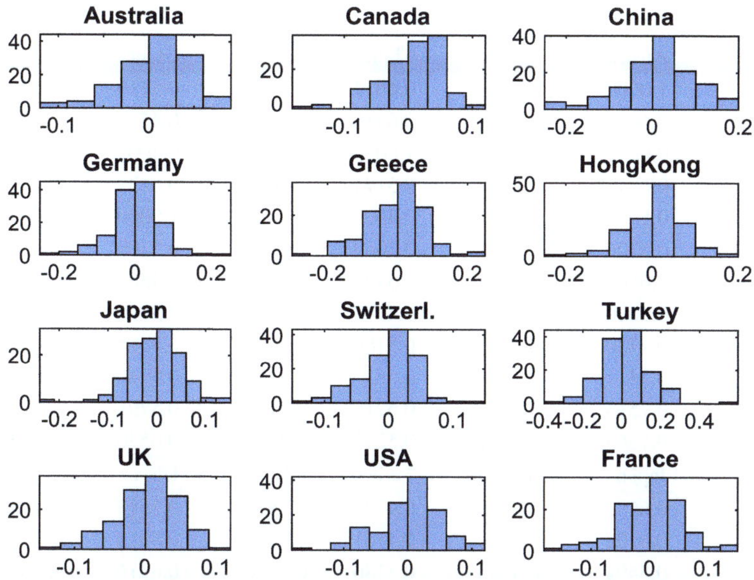

Fig. 3.2 Mean and standard-deviation of given indices

we can confirm our visual assessment as Turkey is the only country with positive skewness in our sample.

```
'Country'          'Skewness'      'Kurtosis'
    'Australia'        [ -0.6285]        [  3.2136]
    'Canada'           [ -0.7792]        [  4.2925]
    'China'            [ -0.4868]        [  3.2715]
    'Germany'          [ -0.4533]        [  4.7077]
    'Greece'           [ -0.1801]        [  3.8300]
    'Hong Kong'        [ -0.2517]        [  4.0294]
    'Japan'            [ -0.4205]        [  4.2201]
    'Switzerl.'        [ -0.4782]        [  3.2599]
    'Turkey'           [  0.5240]        [  5.0382]
    'UK'               [ -0.6239]        [  3.3634]
    'USA'              [ -0.4996]        [  3.5796]
    'France'           [ -0.3844]        [  3.5408]
```

(c)
```
%% c)
CovMat = cov(IndReturn);
CorMat = corrcoef(IndReturn);
% For a nicer display
Result = [ 'Correlation' IndNames ; IndNames'
 num2cell(CorMat)];
```

Table 3.2 Correlation matrix

Corr	Australia	Canada	China	Germany	Greece	Hong Kong
Australia	1.0000	0.6686	0.6039	0.6841	0.5710	0.5706
Canada	0.6686	1.0000	0.6316	0.6570	0.4737	0.7002
China	0.6039	0.6316	1.0000	0.5174	0.4009	0.7069
Germany	0.6841	0.6570	0.5174	1.0000	0.6433	0.6103
Greece	0.5710	0.4737	0.4009	0.6433	1.0000	0.5309
Hong Kong	0.5706	0.7002	0.7069	0.6103	0.5309	1.0000
Japan	0.6563	0.6108	0.5494	0.5556	0.4881	0.5584
Switzerl.	0.6606	0.6162	0.4940	0.7977	0.5950	0.5559
Turkey	0.4720	0.4161	0.5186	0.5770	0.4515	0.4464
UK	0.7411	0.7105	0.5530	0.8124	0.5940	0.6322
USA	0.7513	0.8018	0.6216	0.8177	0.5791	0.6678
France	0.7363	0.7219	0.5235	0.9327	0.6540	0.6113

Corr	Japan	Switzerl.	Turkey	UK	USA	France
Australia	0.6563	0.6606	0.4720	0.7411	0.7513	0.7363
Canada	0.6108	0.6162	0.4161	0.7105	0.8018	0.7219
China	0.5494	0.4940	0.5186	0.5530	0.6216	0.5235
Germany	0.5556	0.7977	0.5770	0.8124	0.8177	0.9327
Greece	0.4881	0.5950	0.4515	0.5940	0.5791	0.6540
Hong Kong	0.5584	0.5559	0.4464	0.6322	0.6678	0.6113
Japan	1.0000	0.5681	0.4583	0.5964	0.6086	0.6018
Switzerl.	0.5681	1.0000	0.4505	0.8147	0.7753	0.8326
Turkey	0.4583	0.4505	1.0000	0.4973	0.5419	0.5878
UK	0.5964	0.8147	0.4973	1.0000	0.8643	0.8788
USA	0.6086	0.7753	0.5419	0.8643	1.0000	0.8413
France	0.6018	0.8326	0.5878	0.8788	0.8413	1.0000

If we look at the correlations given in Table 3.2, we first observe that the correlation between European markets is high. Other expected results are the high co-movements in Chinese and Hong Kong returns, as well as in US, Canadian, Australian and European returns.

(d) The efficient frontier for both the constrained and unconstrained case can be seen in Fig. 3.3.

```
%% d) Compute the efficient frontier
% Without shortsales
% Get global minimum variance portfolio
[weights,mu,varpor] = weightsminc(IndReturn);
% For each mu >= mu of gmvp, get minimum sd in effsigmas
% 3.2 is chosen so that the tangent PF is visible in e)
mus = mu:0.0005:3.2*max(Means);
for i = 1:length(mus)
```

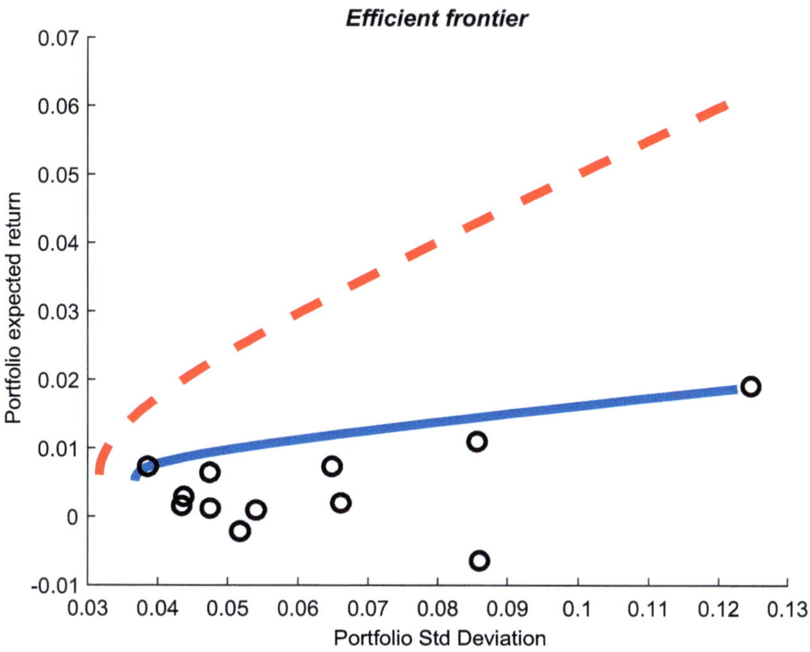

Fig. 3.3 Constrained (solid line) and unconstrained (dashed line) efficient frontiers. Countries are displayed as black dots

```
        if mus(i)>max(Means)
            % Then solver gives wrong solutions since w/o shorts
            % no return greater than max(Means) is possible
        else
            [weights,mu,varpor] = weightsmvc(IndReturn,mus(i));
            effsigmas(i) = varpor^0.5;
            effmus(i) = mu;
        end
    end
    % Plot results
    figure(3)
    hold on
    plot(effsigmas,effmus,'LineWidth', 4);
    ylabel('Portfolio expected return','FontSize',10);
    xlabel('Portfolio Std Deviation','FontSize',10);
    title('\it{Efficient frontier}');
    % Show where the countries lie
    plot(Sigmas,Means,'ko','MarkerSize', 8, 'LineWidth', 2);

    % With Shortsales
```

```
clear effsigmass effmuss;
% Use closed form solution for gmvp
A = ones([1 12])*inv(CovMat)*ones([12 1]);
B = ones([1 12])*inv(CovMat)*Means;
C = Means'*inv(CovMat)*Means;
D = A*C-B^2;
mu = B/A;
varpor = 1/A;


mus = mu:0.0005:3.2*max(Means);

% Calculate the efficient sd for every mu
for i = 1:length(mus)
varpor = (A*mus(i)^2-2*B*mus(i)+C)/D;
effsigmass(i) = varpor^0.5;
effmuss(i) = mus(i);
end
plot(effsigmass,effmuss,'r--','LineWidth', 4);
ylabel('Portfolio expected return','FontSize',10);
xlabel('Portfolio Std Deviation','FontSize',10);
saveas(figure(3),'Frontiers.pdf');
hold off;
```

Appendix for Used Functions: 1. Function File "weightsminc.m"

```
function [weights,mu, varpor]=weightsminc(ret)
% This function returns the weights, expected return and
  variance of the global
% minimum variance portfolio in the constrained case
  (no shortsales)
% calculate the expected returns
[n,col] = size(ret);
expret = (mean(ret))';
% calculate the variance-covariance matrix
expcov = cov(ret);
% set up the quadratic programming
c = zeros(col,1);
% Aeq matrix
Aeq = ones(1,col);
beq = 1;
A = -eye(col,col);
b = -0.0000000001*ones(col,1);
vlb = 0;
vub = 10;
% starting values
```

```
x0 = zeros(col,1);
% quadratic programming function
options = optimset('Algorithm','interior-point-convex')
x = quadprog(expcov, c, A, b, Aeq, beq, vlb, vub,
 x0,options);
%x = quadprog(expcov, c, A, b, Aeq, beq, vlb, vub, x0);
size(x)
% calculate variance portfolio
varpor = x'*expcov*x;
% optimal portfolio weights and the corresponding expected
 return
weights = x;
mu = x'*expret;
end
```

2. Function File "weightsmvc.m"

```
function [weights,mu, varpor]=weightsmvc(ret,expret1)
% This function returns the weights, expected return and
  variance of the efficient
% portfolio with given expected return in the constrained
  case (no shortsales)
[n,col] = size(ret);
expret = (mean(ret))';
% calculate the variance-covariance matrix
expcov = cov(ret);
weights = [];
% set up the quadratic programming
c = zeros(col,1);
% Aeq matrix
Aeq = ones(1,col);
beq = 1;
A=[-eye(col,col); -expret'];
b=[-0.00000001*ones(col,1); -expret1];
vlb = [];
vub = 10;
% starting values
x0 = zeros(col,1);
% quadratic programming function
options = optimset('Algorithm','interior-point-convex')
x = quadprog(expcov, c, A, b, Aeq, beq, vlb, vub, x0,options);
%x = quadprog(expcov, c, A, b, Aeq, beq, vlb, vub, x0);
size(x)
% calculate variance portfolio
varpor = x'*expcov*x;
% optimal portfolio weights and the corresponding expected
 return
```

```
weights = x;
mu=x'*expret;
end
```

(e)
```
%% e) Compute the tangential-portfolio with and without
    short-sales.
% First without shortsales
[col,row] = size(IndReturn);
% Start optimization with equalweight PF
x0 = ones(1,row)*(1/row);
% Shortsales are not allowed
A2 = -eye(row,row);
b = -0.00000001*ones(row,1);
Aeq = ones(1,row);
beq = 1;
lb = [];
ub = [];
% Get tangential PF weights by maximizing sharpe subject
  to constraints
% (In this case minimize negative SR)
lambdaCon = fmincon(@sharpeT,x0',A2,b,Aeq,beq,lb,ub,[],[],
[Rf IndReturn]);
% Get mean and sd, plot
meanT = Means'*lambdaCon;
sigmaT = sqrt(lambdaCon'*CovMat*lambdaCon);
hold on;
plot(sigmaT,meanT,'kx','MarkerSize', 12, 'LineWidth', 2);
% Draw new efficient frontier
Sigs = 0:0.001:max(Sigmas);
for i = 1:length(Sigs)
    EffLine(i) = mean(Rf)+Sigs(i)*(meanT-mean(Rf))/sigmaT;
end
plot(Sigs,EffLine,'k','LineWidth',1);

% Now same thing with shortsales, this time without solver
% Solver would work too, very slightly different results

lambda = inv(CovMat)*(Means-mean(Rf)*ones(length(Means),1))/
(B-A*mean(Rf))
meanTs = Means'*lambda;
sigmaTs = sqrt(lambda'*CovMat*lambda);
plot(sigmaTs,meanTs,'kx','MarkerSize', 12, 'LineWidth', 2);
% Again, draw new efficient frontier, same sigs
for i = 1:length(Sigs)
    EffLine(i) = mean(Rf)+Sigs(i)*(meanTs-mean(Rf))/sigmaTs;
```

```
end
plot(Sigs,EffLine,'k','LineWidth',1);
% Make sure text is not a variable, then add rf label
clear text;
txt1 = 'rf';
text(0.015, -0.005, txt1, 'FontSize',13);
% Annotation placement is relative to full plot window
annotation('arrow', 'X', [0.21, 0.14], 'Y', [0.17, 0.225]);
hold off
saveas(gcf, 'EfficientFrontierwithTangents.pdf');

% Print the weights in a nice way
TangentPortfolios = [{'Countries'}, {'No SS Tangent'} ,
{'Uncon. Tangent'};
    IndNames', num2cell(lambdaCon), num2cell(lambda);
    {'Mean'}, num2cell(meanT), num2cell(meanTs);
    {'SD'}, num2cell(sigmaT), num2cell(sigmaTs)]
```

Figure 3.4 shows the plot of the two tangent portfolios. Much like the two efficient frontiers, they differ greatly. Therefore, much better risk reward tradeoffs are (theoretically) available in the unconstrained case. This is due to the fact, that assets can be sold short for finance investments and therefore small differenced in expected returns can be leveraged. From this also stems the much higher sharpe ratio of the unconstrained tangent portfolio. The weights of the tangent portfolios are:

```
 TangentPortfolios =
'Countries'     'No SS Tangent'     'Uncon. Tangent'
    'Australia'     [       0.7510]     [          3.1623]
    'Canada'        [   4.3154e-05]     [          2.0188]
    'China'         [       0.0051]     [         -0.1757]
    'Germany'       [   2.1266e-06]     [          1.0705]
    'Greece'        [   9.9938e-07]     [         -0.6693]
    'Hong Kong'     [   1.9768e-05]     [          0.4573]
    'Japan'         [   1.7276e-06]     [         -1.6964]
    'Switzerl.'     [   3.3206e-06]     [          0.7502]
    'Turkey'        [       0.2438]     [          0.6510]
    'UK'            [   3.9478e-06]     [          1.3263]
    'USA'           [   2.5177e-06]     [         -3.0256]
    'France'        [   2.1361e-06]     [         -2.8692]
```

(f)
```
%% f)
% Vary sample means by
delta = 0.01;
amount = 5;
```

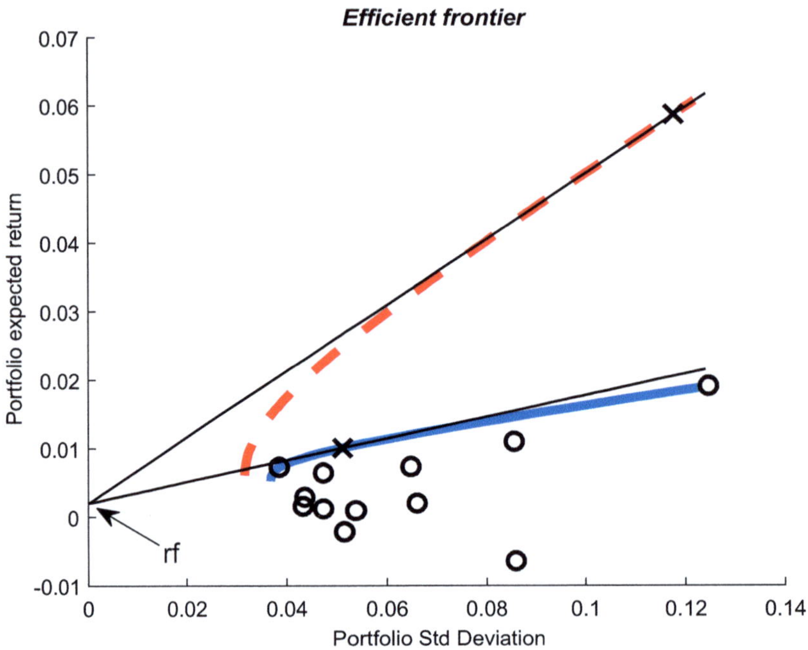

Fig. 3.4 Constrained (solid line) and unconstrained (dashed line) efficient frontiers with no riskless asset and the respective tangent portfolios (X). The new efficient frontiers including the riskless asset (tangential portfolios) is displayed as thin solid lines. Countries are shown as black dots

```
means = Means;
lambda = ones(length(means),amount);
% To plot
figure(4)
hold on
linS = {'-','--',':', '-.', ':'};
mS = {'none', 'none', 'none', 'none', '^'};
% Define handles for both plots, this is needed for legends
H1 = gobjects(5,1);
H2 = gobjects(5,1);

for j = 1:amount
    clear effsigmas effmus;
    % Use closed form solution for gmvp
    A = ones([1 12])*inv(CovMat)*ones([12 1]);
    B = ones([1 12])*inv(CovMat)*means;
    C = means'*inv(CovMat)*means;
    D = A*C-B^2;
    mu = B/A;
```

```
    gmv = 1/A;
    % Calculate the efficient sd for every mu
    mus = mu:0.001:3.2*max(Means);
    for i = 1:length(mus)
        varpor = (A*mus(i)^2-2*B*mus(i)+C)/D;
        effsigmas(i) = varpor^0.5;
        effmus(i) = mus(i);
    end
    % For each frontier, calculate tangent PF
    lambda = inv(CovMat)*(means-mean(Rf)*ones(length
    (means),1))/(B-A*mean(Rf));
    meanTs = means'*lambda;
    sigmaTs = sqrt(lambda'*CovMat*lambda);
    % Go to next mean step
    means = means+delta;
    figure(4)
    H1(j) = plot(effsigmas,effmus,'LineWidth', 2,
     'LineStyle', linS{j}, 'Marker', mS{j},
    'MarkerIndices',1:4:length(effmus));
    plot(sigmaTs,meanTs,'k*','MarkerSize', 12,
     'LineWidth', 2);
    xlabel('Std. Deviation');
    ylabel('Expected Return');
    % Show change in weights
    figure(5)
    hold on;
    xlabel('Assets');
    ylabel('Weights in tangent PF');
    H2(j) = plot(lambda,'LineWidth', 2, 'LineStyle',
     linS{j}, 'Marker', mS{j});
    legtext{j} = ['$\delta = ', num2str((j-1)*delta), '$'];
end
% Draw the legends and save figures
leg1 = legend(H1, legtext, 'FontSize', 11, 'Location',
 'southeast');
leg2 = legend(H2, legtext, 'FontSize', 11, 'Location',
 'southwest');
set(leg1,'Interpreter','latex');
set(leg2,'Interpreter','latex');
hold off
saveas(4,'EfficientFrontierandTangentPortsSensitivity
1.pdf');
saveas(5,'EfficientFrontierandTangentPortsSensitivity
2.pdf');
```

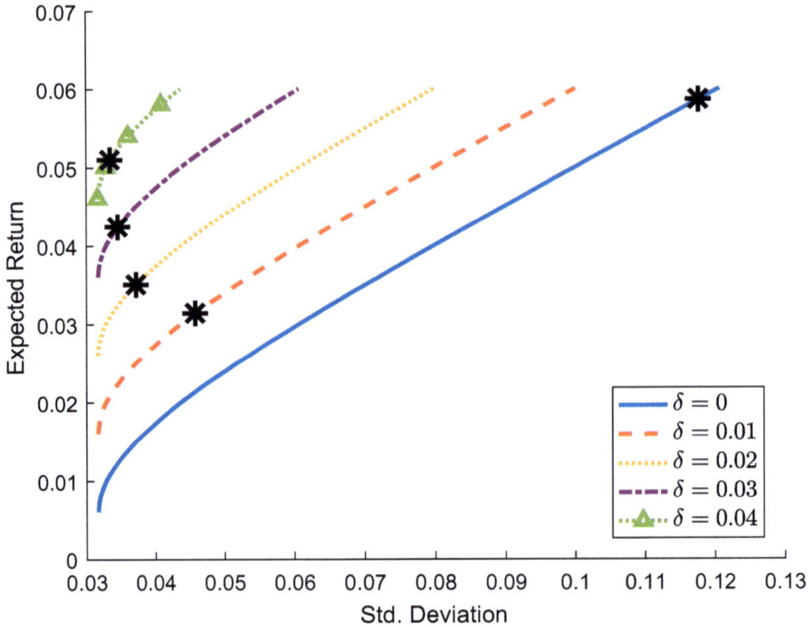

Fig. 3.5 Sensitivity of efficient frontier and tangent portfolio (*) with regards to changes in means

Figure 3.5 shows how sensitive the efficient frontier and tangent portfolio react to changes in expected returns. We plotted the original efficient frontier and four others, each with expected returns of all assets increased by one percentage point. The strong effect of changes in expected returns can also be seen in Fig. 3.6, which shows the weights of the tangent portfolio for the same set of expected return deviations. We conclude that the efficient frontier and tangent portfolio are highly sensitive to expected returns and thus small errors in estimation can result in large deviations.

(g) The security market line is shown in Fig. 3.7. In this example, all assets are on the SML by construction.

```
%% g) Plot the SML
clear lambda;
% Calculate tangential PF with shortsales as before
% Prepare variables again since they were varied before
A=ones([1 12])*inv(CovMat)*ones([12 1]);
B=ones([1 12])*inv(CovMat)*Means;
C=Means'*inv(CovMat)*Means;
D=A*C-B^2;
mu=B/A;
varpor=1/A;
```

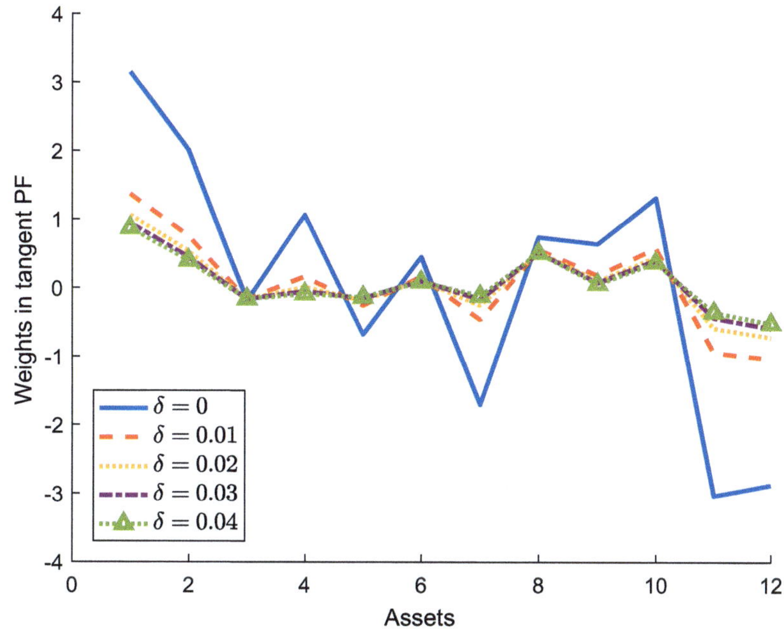

Fig. 3.6 Sensitivity of efficient PF weights

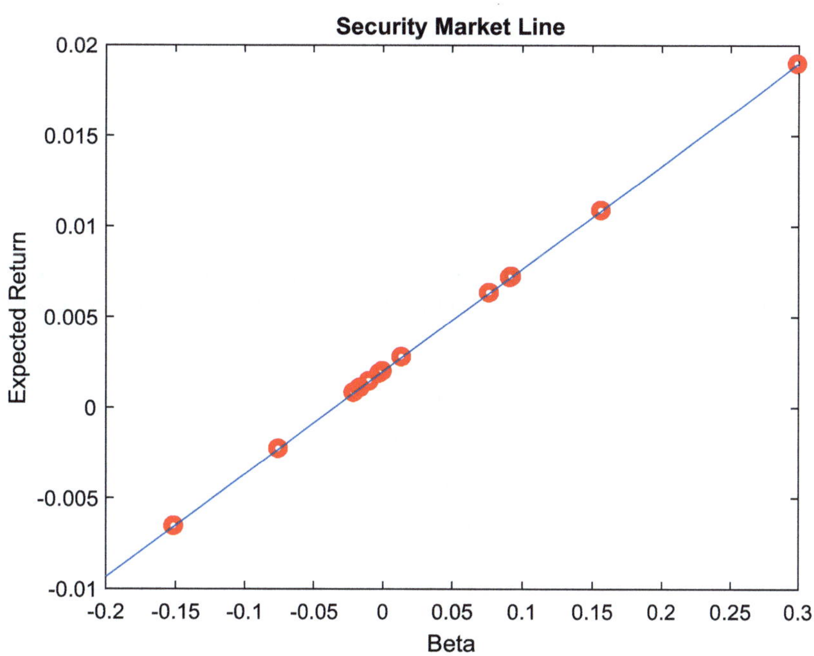

Fig. 3.7 Security market line

```
lambda=inv(CovMat)*(Means-mean(Rf)*ones(length(Means),1))/
(B-A*mean(Rf));
market=IndReturn*lambda;

% Excess return of market (=tangent portfolio)
excessmarket=market-Rf;

[row,col]=size(IndReturn);
% Calculate betas for each asset
for j=1:col
y=IndReturn(:,j)-Rf;
b(j)=regress(y,excessmarket);
end

% Plot the assets
figure(6)
plot([0 b'],[mean(Rf),Means'],'ro','LineWidth',3)
hold on
%plot([0 1],[mean(Rf),mean(market)]);

% Plot the SML
xlim = get(gca,'XLim');
y1 = mean(Rf) + xlim(1)*(mean(excessmarket));
y2 = mean(Rf) + xlim(2)*(mean(excessmarket));
line([xlim(1) xlim(2)],[y1 y2]);
title('Security Market Line')
xlabel('Beta')
ylabel('Expected Return')
saveas(6,'SML.pdf');
```

3.3 Consider, e.g., an economy with three equally likely states and with the following assets returns: $R^1 = (1\%, 0\%, 0\%)'$, $R^2 = (0\%, 1\%, 0\%)'$ and $R^3 = (0\%, 0\%, 1\%)'$. Then

$$\begin{aligned} corr(R^1, R^2) &= \frac{cov(R^1, R^2)}{\sigma(R^1)\sigma(R^2)} = \frac{\mathbb{E}(R^1 R^2) - \mathbb{E}(R^1)\mathbb{E}(R^2)}{\sigma(R^1)\sigma(R^2)} \\ &= -\frac{1/3 \cdot 1/3}{\sqrt{2}/3 \cdot \sqrt{2}/3} = -\frac{1}{2}, \end{aligned}$$

similarly, $corr(R^1, R^3) = corr(R^2, R^3) = -\frac{1}{2}$. On the other hand,

$$corr((-R^1 - R^2), R^3) = \frac{\mathbb{E}((-R^1 - R^2)R^3) - \mathbb{E}(-R^1 - R^2)\mathbb{E}(R^3)}{\sigma(-R^1 - R^2)\sigma(R^3)}$$
$$= \frac{2/3 \cdot 1/3}{\sqrt{2}/3 \cdot \sqrt{2}/3} = 1.$$

As discussed in the book, in presence of two assets with correlation of 1 or -1, usually the tangent portfolio cannot be defined, since for a given level of variance, a portfolio with an arbitrarily large return can be constructed. In the example above we have shown that no pair of assets with correlation of 1 or -1 guarantees the existence of the tangent portfolio, as some combination of two assets still might correlate with a third asset equal to 1 or -1. Therefore, in order to guarantee the existence of the tangent portfolio, one has to exclude the presence of two portfolios with correlation between their returns equal to 1 or -1, see Theorem 3.4 on page 102 of the book.

3.4

(a) The optimal portfolio λ^{opt} can be found as a solution of the following decision problem of the investor:

$$\lambda^{opt} = \arg\max_{\lambda \in \mathbb{R}^4} U(\mu_\lambda, \sigma_\lambda^2) = \mu_\lambda - \sigma_\lambda^2,$$
$$\text{s.t.} \sum_{k=0}^{3} \lambda_k = 1,\ 0 \le \lambda_1,\ 0 \le \lambda_2,\ \lambda_3 = 0. \qquad (3.7)$$

Define

$$\bar{\mu} = (\mu_1, \mu_2)^T, \quad \overline{COV} = \begin{pmatrix} \sigma_1^2 & \sigma_{1,2}^2 \\ \sigma_{2,1}^2 & \sigma_2^2 \end{pmatrix}.$$

Plugging the budget constraint $\lambda_0 = 1 - \sum_{k=1}^{3} \lambda_k$, we can rewrite (3.7) as follows:

$$\begin{pmatrix} \lambda_1^{opt} \\ \lambda_2^{opt} \end{pmatrix} = \arg\max_{\lambda \in \mathbb{R}^2} (\bar{\mu} - R_f \mathbf{1})^T \lambda - \lambda^T \overline{COV} \lambda, \ \text{s.t. } 0 \le \lambda_1, 0 \le \lambda_2.$$

The unconstrained problem above has already been solved in 3.1(a):

$$\lambda_1^{opt} = \frac{5}{2\rho} = 1.25,$$

$$\lambda_2^{opt} = \frac{2}{\rho} = 1,$$

where $\rho = 2$. As for remaining components, $\lambda_3^{opt} = 0$ and $\lambda_0^{opt} = 1 - \sum_{k=1}^{3} \lambda_k^{opt} = -1.25$.

Finally,

$$\mu_{\lambda^{opt}} = \lambda_0^{opt} R_f + \lambda_1^{opt} \mu_1 + \lambda_2^{opt} \mu_2 = 11.25\%,$$

$$\sigma^2_{\lambda^{opt}} = (\lambda_1^{opt})^2 \sigma_1^2 + 2\lambda_1^{opt} \lambda_2^{opt} \sigma_{1,2}^2 + (\lambda_2^{opt})^2 \sigma_2^2 = 4.63\%,$$

$$U(\mu_{\lambda^{opt}}, \sigma^2_{\lambda^{opt}}) = \mu_{\lambda^{opt}} - \sigma^2_{\lambda^{opt}} = 6.62\%.$$

(b) By definition,

$$\alpha_{\lambda,k} = (\mu_k - R_f) - \rho \cdot cov(R_k, \sum_{l=1}^{K} R_l \lambda_l), \quad k = 1, 2, 3. \tag{3.8}$$

In our example, $\rho = 2$, therefore,

$$\alpha_{\lambda^{opt},1} = (\mu_1 - R_f) - 2 \sum_{l=1}^{3} \lambda_l^{opt} \sigma_{1,l}^2 = 0\%,$$

$$\alpha_{\lambda^{opt},2} = (\mu_2 - R_f) - 2 \sum_{l=1}^{3} \lambda_l^{opt} \sigma_{2,l}^2 = 0\%,$$

$$\alpha_{\lambda^{opt},3} = (\mu_3 - R_f) - 2 \sum_{l=1}^{3} \lambda_l^{opt} \sigma_{3,l}^2 = 1\%.$$

Now suppose the investor creates a new portfolio λ^* investing, e.g., a fraction $\gamma = 0.9$ of his wealth into the optimal portfolio λ^{opt} and $(1 - \gamma) = 0.1$ into the asset 3. Then

$$\mu_{\lambda^*} = \gamma \cdot \mu_{\lambda^{opt}} + (1 - \gamma) \cdot \mu_3 = 11.125\%,$$

$$\sigma^2_{\lambda^*} = \gamma^2 \cdot \sigma^2_{\lambda^{opt}} + 2 \cdot \gamma \cdot (1 - \gamma) \cdot cov(\lambda^{opt}, R_3) + (1 - \gamma)^2 \cdot \sigma_3^2 = 4.460\%,$$

$$U(\mu_{\lambda^*}, \sigma^2_{\lambda^*}) = \mu_{\lambda^*} - \sigma^2_{\lambda^*} = 6.665\% > U(\mu_{\lambda^{opt}}, \sigma_{\lambda^{opt}}),$$

where we used that

$$cov(\lambda^{opt}, R_3) = \lambda_1^{opt}\sigma_{1,3}^2 + \lambda_2^{opt}\sigma_{2,3}^2 = 3.5\%.$$

(c) Denote by λ^{γ} the investment strategy which allocates a fraction γ of wealth into the asset 2, while the remaining portion, $1-\gamma$, into the asset 3. Then

$$\begin{aligned}
U(\mu_{\lambda^{\gamma}}, \sigma_{\lambda^{\gamma}}^2) - U(\mu_2, \sigma_2^2) &= (\mu_{\lambda^{\gamma}} - \mu_2) - (\sigma_{\lambda^{\gamma}}^2 - \sigma_2^2) \\
&= (1-\gamma)(\mu_3 - \mu_2) - ((\gamma^2 - 1)\sigma_2^2 + 2\gamma(1-\gamma)\sigma_{2,3}^2 + (1-\gamma)^2\sigma_3^2) \\
&= (1-\gamma)\cdot 2.5\% - ((\gamma^2 - 1)\cdot 4\% + 2\gamma(1-\gamma)\cdot 6\% + (1-\gamma)^2 \cdot 8\%) \\
&= -(1-\gamma)\cdot 1.5\%.
\end{aligned}$$

If no short selling is allowed, i.e., $0 \le \gamma \le 1$, then $U(\mu_{\lambda^{\gamma}}, \sigma_{\lambda^{\gamma}}^2) - U(\mu_2, \sigma_2^2) \le 0$, hence adding asset 3 to the portfolio consisting of asset 2 only leaves the investor worse off.

3.5

(a) Recall the expression under section "Active or Passive?" on page 124 of the book: an agent should be active if and only if:

$$U_{\hat{\mu}}^i(\mu^i) - U_{\hat{\mu}}^i(\bar{\mu}) = \frac{1}{2\rho^i}\left(||\hat{\mu} - \bar{\mu}||^2 - ||\hat{\mu} - \mu^i||^2\right) > C^i$$

where $C^i > 0$ when $\mu^i - \bar{\mu} \neq 0$ and $||x||^2 := x'COV^{-1}x$.

$$\frac{1}{2\rho^i}\left(\underbrace{||\hat{\mu} - \bar{\mu}||^2}_{\text{Market Inefficiency Term}} - \underbrace{||\hat{\mu} - \mu^i||^2}_{\text{Individual beliefs' deviation}}\right) > C^i$$

Active:

- the less efficient the market
- the more skilled the investor
- the smaller his cost to be active
- the less risk averse she is.

$$||\hat{\mu} - \bar{\mu}||^2 = \frac{25}{32} = 0.781$$

$$\frac{1}{2\rho^i}\left(||\hat{\mu} - \bar{\mu}||^2 - ||\hat{\mu} - \mu^i||^2\right)$$

Investor 1:

$$\frac{1}{4}(0.781 - 8.5) < 0,$$

this is very negative, so as C^1 is positive, it will be lower then C^1.

Investor 2:

$$\frac{1}{4}(0.781 - 0.5) > 0,$$

Investor 3:

$$\frac{1}{4}(0.781 - 0.5) > 0,$$

Investor 4:

$$\frac{1}{4}(0.781 - 5) < 0,$$

If the investor 2 and 3 do not have high costs of acquiring information C^i, then it will be optimal for them to be active and investor 1 and 4 should rather be passive.

Active portfolios of all individuals are calculated by using the standard mean-variance optimization' first order condition:

Recall the first order condition:

$$COV\lambda^i = \frac{1}{\rho^i}\left(\mu^i - R_f\mathbb{I}\right)$$

solving it for every individual according to their belief of the expected returns of the assets, we have:

$$\lambda^1 = \begin{bmatrix} 5/4 \\ 0 \end{bmatrix},\ \lambda^2 = \begin{bmatrix} 1/2 \\ 1/4 \end{bmatrix},\ \lambda^3 = \begin{bmatrix} 1/4 \\ 1/2 \end{bmatrix},\ \text{and } \lambda^4 = \begin{bmatrix} 0 \\ 1 \end{bmatrix}$$

Then the only risky asset portfolios are attained by $\lambda^i / \sum_k \lambda^i_k$:

$$\bar{\lambda}^1 = \begin{bmatrix} 1 \\ 0 \end{bmatrix},\ \bar{\lambda}^2 = \begin{bmatrix} 2/3 \\ 1/3 \end{bmatrix},\ \bar{\lambda}^3 = \begin{bmatrix} 1/3 \\ 2/3 \end{bmatrix},\ \text{and } \bar{\lambda}^4 = \begin{bmatrix} 0 \\ 1 \end{bmatrix}$$

The market portfolio weights are then:

$$\bar{\lambda}^M = \sum_{i=1}^{4} r^i \bar{\lambda}^i = \begin{bmatrix} \frac{8}{15} \\ \frac{7}{15} \end{bmatrix}$$

where the relative wealth invested in financial markets r^i is calculated according the total wealth invested in the financial markets $(1-\lambda_0)W^i$, hence:

$$r^i = \frac{(1-\lambda_0^i)W^i}{\sum_{i=1}^4 (1-\lambda_0^i)W^i}$$

we find $r^1 = 1/3, r^2 = r^3 = 1/5, r^4 = 4/15$

The beta of the two assets are:

$$\beta = \frac{COV\lambda^M}{(\lambda^M)'COV\lambda^M} = \begin{bmatrix} \frac{120}{113} \\ \frac{105}{113} \end{bmatrix}$$

The true expected return of the market portfolio is:

$$\hat{\mu}^M = \begin{bmatrix} \frac{8}{15} & \frac{7}{15} \end{bmatrix} \begin{bmatrix} 2 \\ 2 \end{bmatrix} = 2$$

Then the CAPM gives us

$$\hat{\alpha} = \hat{\mu} - R_f \mathbb{I} - \beta \left(\hat{\mu}^M - R_f \right)$$

$$\hat{\alpha} = \begin{bmatrix} 2 \\ 2 \end{bmatrix} - \begin{bmatrix} 1 \\ 1 \end{bmatrix} - \begin{bmatrix} \frac{120}{113} \\ \frac{105}{113} \end{bmatrix} (2-1) = \begin{bmatrix} \frac{-7}{113} \\ \frac{8}{113} \end{bmatrix}.$$

All investors alpha's then:

$$\hat{\alpha}^1 = \hat{\alpha}'\bar{\lambda}^1 = \frac{-7}{113}$$

$$\hat{\alpha}^2 = \hat{\alpha}'\bar{\lambda}^2 = \frac{-2}{113}$$

$$\hat{\alpha}^3 = \hat{\alpha}'\bar{\lambda}^3 = \frac{3}{113}$$

$$\hat{\alpha}^4 = \hat{\alpha}'\bar{\lambda}^4 = \frac{8}{113}$$

(b)

$$\hat{\alpha}^2 = \hat{\alpha}'\bar{\lambda}^2 = \frac{-2}{113}$$

however, as investor 2:

$$\frac{1}{4}(0.781 - 0.5) > 0,$$

investor 2 should be rather active.

(c)

$$\hat{\alpha}^4 = \hat{\alpha}'\bar{\lambda}^4 = \frac{8}{113}$$

however, as investor 4:

$$\frac{1}{4}(0.781 - 5) < 0,$$

she should rather be passive although she could have a positive alpha portfolio.

3.6

(a) Remember from the script that the ex-post alpha of asset k is:

$$\hat{\alpha}_{k,M} = \left(\hat{\mu}_k - R_f\right) - \beta_{k,M}\left(\hat{\mu}_M - R_f\right),$$

where $\hat{\mu}_k$ the true expected return of asset k and $\hat{\mu}_M$ the true expected return of the market portfolio. The ex-post alpha of the strategy λ_k^i is then:

$$\hat{\alpha}_M^i = \sum_{k=1}^{K} \lambda_k^i \hat{\alpha}_{k,M}.$$

(b) To show that the market is a zero sum game we have to prove that $\sum_i r^i \hat{\alpha}_M^i = 0$ (r^i is the fraction of the initial endowment investor i got). This is done below:

$$\begin{aligned}\sum_i r^i \hat{\alpha}_M^i &= \sum_i r^i \sum_k \lambda_k^i \hat{\alpha}_{k,M} \\ &= \sum_k \lambda_k^M \left[\left(\hat{\mu}_k - R_f\right) - \beta_{k,M}\left(\hat{\mu}_M - R_f\right)\right],\end{aligned}$$

since $\lambda_0^M = 0$ and $\beta_M = 1$

$$\sum_i r^i \hat{\alpha}_M^i = \left(\hat{\mu}_M - R_f\right) - \beta_M\left(\hat{\mu}_M - R_f\right) = 0.$$

(c) The CAPM tells us that

$$(\mu_k - R_f) = \beta_k(\mu_M - R_f),$$

In other words, with a high enough beta, every average return can be explained. The problem with Hedge Funds (HF) is that they have quite a low beta but large average returns. At this point, the classical CAPM gets into trouble.

In the CAPM with heterogeneous beliefs, high HF returns can be explained by the assumption that HF managers are well informed. But in an equilibrium, this effect should be washed out (because all uninformed investors invest passively).

(d) The zero sum game has been proved before. Furthermore, the statement "there is no theory which describes high HF returns" is not true anymore, since the CAPM with heterogenous beliefs does just that.

A last comment on the high returns and the low betas of HF: from an empirical perspective, these facts are not as clear as claimed by HF marketing.

3.7

(a) The market expectation is

$$\bar{\mu} = \sum_i \frac{\frac{r^i}{\gamma^i}}{\sum_j \frac{r^j}{\gamma^j}} \mu^i = \sum_i \frac{1}{3}\mu^i = \begin{pmatrix} 2 \\ \frac{4}{3} \end{pmatrix}.$$

(b) From the first order condition of the utility maximization problem we know the optimal portfolio of investor i:

$$\lambda_i^* = \frac{1}{\gamma^i} \operatorname{cov}^{-1}\left(\mu^i - R_f\right),$$

From that it we obtain:

$$\lambda_1^* = \frac{1}{2}\begin{pmatrix} 1 \\ 0 \end{pmatrix}, \qquad \lambda_{0,1}^* = \frac{1}{2},$$

$$\lambda_2^* = \frac{1}{4}\begin{pmatrix} 1 \\ 0 \end{pmatrix}, \qquad \lambda_{0,2}^* = \frac{3}{4},$$

$$\lambda_3^* = \frac{1}{4}\begin{pmatrix} 0 \\ 1 \end{pmatrix}, \qquad \lambda_{0,3}^* = \frac{3}{4}.$$

(c) The market clearing condition is

$$\lambda_M \left(\sum_j (1 - \lambda^*_{0,j}) w_0^j \right) = \sum_i w_0^i \lambda^*_i .$$

From that the market portfolio is obtained:

$$\lambda_M = \sum_i \frac{w_0^i}{\sum_j (1 - \lambda^*_{0,j}) w_0^j} \lambda^*_i = \frac{1}{4} \begin{pmatrix} 3 \\ 1 \end{pmatrix}.$$

(d) From the script it is known that the difference between the utility of investing active and investing passive can be written as:

$$U^i_{active} - U^i_{passive} = \frac{1}{2\gamma^i} \left(\|\hat{\mu} - \bar{\mu}\|^2 - \left\|\hat{\mu} - \mu^i\right\|^2 \right),$$

where $\|x\|^2 := x' \operatorname{cov}^{-1} x$. Therefore we get:

$$\|\hat{\mu} - \bar{\mu}\|^2 = \frac{1}{18} \quad \left\|\hat{\mu} - \mu^1\right\|^2 = \frac{1}{2} \quad \left\|\hat{\mu} - \mu^2\right\|^2 = 0 \quad \left\|\hat{\mu} - \mu^3\right\|^2 = 1.$$

Therefore, we can conclude that investor 1 and 3 should invest passive and investor 2 should invest active.

(e) To calculate $\hat{\alpha}_M$ we need $\hat{\mu}_M$ and β:

$$\hat{\mu}_M = \lambda^{M\,\prime} \hat{\mu} = \frac{7}{4},$$

$$\beta = \frac{\operatorname{cov}(R, R_M)}{\sigma_M^2} = \frac{\operatorname{cov} \lambda_M}{\lambda'_M \operatorname{cov} \lambda_M} = \frac{2}{5} \begin{pmatrix} 3 \\ 1 \end{pmatrix},$$

$\hat{\alpha}_M$ is then

$$\hat{\alpha}_M = (\hat{\mu} - R_f) - \beta\,(\hat{\mu}_M - R_f) = \frac{1}{10} \begin{pmatrix} 1 \\ -3 \end{pmatrix}.$$

The alphas of the investors' portfolios are then

$$\hat{\alpha}^1_M = \lambda^{*\,\prime}_1 \hat{\alpha}_M = \frac{1}{20} \qquad \hat{\alpha}^2_M = \frac{1}{40} \qquad \hat{\alpha}^3_M = -\frac{3}{40}.$$

(f) The ex-post alpha of investor 1 is $\frac{1}{20}$, if he invests active. But we know from before that investor 1 has a higher utility, if he invests passive.

3.8

(a) The tangent portfolio is the portfolio with maximum Sharpe ratio i.e. the portfolio λ^T that satisfies:

$$\lambda^T = argmax_{\lambda} \frac{\mu_\lambda}{\sigma_\lambda}$$

s.t.

$$\lambda_1^T, \lambda_2^T \geq 0,$$
$$\lambda_1^T + \lambda_2^T = 1.$$

An easier way to find the tangent portfolio is to follow the two fund separation theory, and optimize the utility of a mean-variance investor, i.e. to find the λ^i that solves the following problem:

$$max_{\lambda^i}\ U^i = \mu_\lambda - \frac{\rho^i}{2}\sigma_\lambda$$

s.t.

$$\lambda_0^i, \lambda_1^i, \lambda_2^i \geq 0,$$
$$\lambda_0 + \lambda_1 + \lambda_2 = 1.$$

and then to normalize it to get the shares in a portfolio of only risky assets, which according to the two fund separation theorem coincides with the tangent portfolio. If there are no constraints on λ^i, then by substituting $\lambda_0^i = 1-(\lambda_1^i+\lambda_2^i)$ the above stated problem can be simplified to:

$$\lambda^i = \frac{1}{\rho^i}\begin{bmatrix} 5 & 25 \\ 25 & 10 \end{bmatrix}^{-1}\left(\begin{bmatrix} 5 \\ 10 \end{bmatrix} - \begin{bmatrix} 0 \\ 0 \end{bmatrix}\right) = \begin{bmatrix} \frac{0.35}{\rho^i} \\ \frac{0.13}{\rho^i} \end{bmatrix}.$$

So λ_1^i and λ_2^i are both positive.

The tangent portfolio is thus obtained as:

$$\lambda_1^T = \frac{\lambda_1^i}{\lambda_1^i + \lambda_2^i} = \frac{\frac{0.35}{\rho^i}}{\frac{0.48}{\rho^i}} = 0.73,$$

$$\lambda_2^T = \frac{\lambda_2^i}{\lambda_1^i + \lambda_2^i} = \frac{\frac{0.13}{\rho^i}}{\frac{0.48}{\rho^i}} = 0.27.$$

(b) As stated above the shares of tangent portfolio equal to:

$$\lambda^T = \frac{\frac{1}{\rho^i}\left(COV^{-1}(\mu - R_f\mathbb{1})\right)}{\frac{1}{\rho^i}\left(COV^{-1}(\mu - R_f\mathbb{1})\right)_1 + \frac{1}{\rho^i}\left(COV^{-1}(\mu - R_f\mathbb{1})\right)_2}$$

$$= \frac{\begin{bmatrix} -0.017 & 0.043 \\ 0.043 & -0.0087 \end{bmatrix}\begin{bmatrix} 5 - R_f \\ 10 - R_f \end{bmatrix}}{\left(\begin{bmatrix} -0.017 & 0.043 \\ 0.043 & -0.0087 \end{bmatrix}\begin{bmatrix} 5 - R_f \\ 10 - R_f \end{bmatrix}\right)_1 + \left(\begin{bmatrix} -0.017 & 0.043 \\ 0.043 & -0.0087 \end{bmatrix}\begin{bmatrix} 5 - R_f \\ 10 - R_f \end{bmatrix}\right)_2}$$

$$= \frac{\begin{bmatrix} -0.017(5 - R_f) + 0.043(10 - R_f) \\ 0.043(5 - R_f) - 0.0087(10 - R_f) \end{bmatrix}}{\begin{bmatrix} -0.017(5 - R_f) + 0.043(10 - R_f) \\ 0.043(5 - R_f) - 0.0087(10 - R_f) \end{bmatrix}' \mathbb{1}} = \begin{bmatrix} \frac{0.345 - 0.026R_f}{0.473 - 0.0603R_f} \\ \frac{0.128 - 0.0343R_f}{0.473 - 0.0603R_f} \end{bmatrix}.$$

Therefore, the derivative with respect to the risk-free rate, evaluated at equals:

$$\frac{\partial \lambda^T}{\partial R_f} = \begin{bmatrix} \frac{-0.026(0.0473 - 0.0603R_f) - 0.0603(0.345 - 0.026R_f)}{(0.473 - 0.0603R_f)^2} \\ \frac{-0.0343(0.473 - 0.0603R_f) - 0.0603(0.128 - 0.0343R_f)}{(0.473 - 0.0603R_f)^2} \end{bmatrix}$$

$$= \begin{bmatrix} \frac{-0.033}{0.224} \\ \frac{-0.024}{0.224} \end{bmatrix} = \begin{bmatrix} -0.15 \\ -0.11 \end{bmatrix}.$$

This is also intuitive, since with the increase of the risk-free rate, investing in risk-free asset becomes more attractive relative to other assets.

4 Two-Period Model: State-Preference Approach

4.1

(a) The budget sets are:
  (A) Arrow-Debreu equilibrium

$$\sum_{s=0}^{\mathbb{S}} \pi_s c_s^i \leq \sum_{s=0}^{\mathbb{S}} \pi_s \omega_s^i \underbrace{\Rightarrow}_{\text{Normalizing } \pi_0=1} c_0^i + \sum_{s=1}^{\mathbb{S}} \pi_s c_s^i \leq \omega_0^i + \sum_{s=1}^{\mathbb{S}} \pi_s \omega_s^i .$$

  (B) Financial Market equilibrium

$$c_0^i + \sum_{k=0}^{K} q^k \theta^{i,k} = \omega_0^i,$$

$$c_s^i = \sum_{k=0}^{K} A_s^k \theta^{i,k} + \omega_s^i .$$

(b) The market equilibria are:
  (A) Every agent allocates his consumption $(c_0^i, c_1^i, \ldots, c_S^i)$ such that

T. Hens, M. O. Rieger, *Solutions to* Financial Economics,
Springer Texts in Business and Economics,
https://doi.org/10.1007/978-3-662-59889-4_12

– Every agent maximizes his own utility given the budget constraints

$$(c_0^i, c_1^i, \ldots, c_S^i) \in \arg\max_{\mathbf{c^i}} \ \mathbf{U}^i(\mathbf{c}^i) \quad \text{s.t} \quad c_0^i + \sum_{s=1}^{\mathbb{S}} \pi_s c_s^i \leq \omega_0^i + \sum_{s=1}^{\mathbb{S}} \pi_s \omega_s^i.$$

– Markets clear

$$\sum_{i=1}^{I} c_s^i = \sum_{i=1}^{I} w_s^i, \forall s = 0, 1, \ldots .S.$$

(B) For the Financial Market equilibrium, we have to specify the initial aggregate supply of the financial assets. If assets are in zero net supply or in unit net supply, the market clearing condition will change according to this.

– Every agent maximizes his own utility given the budget constraints

$$(c_0^i, c_1^i, \ldots, c_S^i) \in \arg\max_{\mathbf{c^i}} \ \mathbf{U}^i(\mathbf{c}^i) \quad \text{s.t} \quad c_0^i + \sum_{k=0}^{K} q^k \theta^{i,k} = \omega_0^i,$$

$$c_s^i = \sum_{k=0}^{K} A_s^k \theta^{i,k} + \omega_s^i.$$

– Markets clear

$$\sum_{i=1}^{I} c_s^i = \sum_{i=1}^{I} w_s^i, \forall s = 0, 1, \ldots .S,$$

and zero net supply case for asset markets clearing $\Rightarrow$

$$\sum_{i=1}^{I} \theta^{i,k} = 0, \forall k = 0, 1, \ldots .S,$$

Or unit net supply case $\Rightarrow$

$$\sum_{i=1}^{I} \hat{\theta}^{i,k} = \sum_{i=1}^{I} \theta_A^{i,k} = 1, \forall k = 0, 1, \ldots .S.$$

(c) Recall the budget constraint of the financial markets equilibrium:

$$c_0^i = \omega_0^i - \sum_{k=0}^{K} q^k \theta^{i,k},$$

$$c_s^i = \sum_{k=0}^{K} A_s^k \theta^{i,k} + \omega_s^i.$$

In the matrix notation, we have

$$\begin{bmatrix} c_0^i \\ c_1^i \\ \cdot \\ \cdot \\ \cdot \\ c_S^i \end{bmatrix} - \begin{bmatrix} \omega_0^i \\ \omega_1^i \\ \cdot \\ \cdot \\ \cdot \\ \omega_S^i \end{bmatrix} = \begin{bmatrix} -q_1 & -q_2 & .. & -q_K \\ A_{11} & A_{12} & .. & A_{1K} \\ \cdot & \cdot & .. & \cdot \\ \cdot & \cdot & .. & \cdot \\ \cdot & \cdot & .. & \cdot \\ A_{S1} & A_{S2} & .. & A_{SK} \end{bmatrix} \begin{bmatrix} \theta_1 \\ \theta_1 \\ \cdot \\ \cdot \\ \cdot \\ \theta_K \end{bmatrix}.$$

We assume markets are arbitrage-free, and complete, then we have the strictly positive linear pricing rule π_s and let us denote the $\bar{\pi} = [1 \ \pi_1 \ \pi_2 . . . \pi_S]'$.
If we multiply both sides of the budget constraint with $\bar{\pi}'$ we get:

$$\bar{\pi}' \begin{bmatrix} c_0^i \\ c_1^i \\ \cdot \\ \cdot \\ \cdot \\ c_S^i \end{bmatrix} - \bar{\pi}' \begin{bmatrix} \omega_0^i \\ \omega_1^i \\ \cdot \\ \cdot \\ \cdot \\ \omega_S^i \end{bmatrix} = \bar{\pi}' \begin{bmatrix} -q_1 & -q_2 & .. & -q_K \\ A_{11} & A_{12} & .. & A_{1K} \\ \cdot & \cdot & .. & \cdot \\ \cdot & \cdot & .. & \cdot \\ \cdot & \cdot & .. & \cdot \\ A_{S1} & A_{S2} & .. & A_{SK} \end{bmatrix} \begin{bmatrix} \theta_1 \\ \theta_1 \\ \cdot \\ \cdot \\ \cdot \\ \theta_K \end{bmatrix}$$

$$= \begin{bmatrix} \underbrace{-q_1 + \sum_{s=1}^{S} \pi_s A_{s1}}_{=0} & \underbrace{-q_2 + \sum_{s=1}^{S} \pi_s A_{s2}}_{=0} & .. & \underbrace{-q_K + \sum_{s=1}^{S} \pi_s A_{sK}}_{=0} \end{bmatrix} \begin{bmatrix} \theta_1 \\ \theta_1 \\ \cdot \\ \cdot \\ \cdot \\ \theta_K \end{bmatrix} = 0.$$

This yields the Arrow Debreu economy budget constraints:

$$\bar{\pi}' \begin{bmatrix} c_0^i \\ c_1^i \\ \cdot \\ \cdot \\ \cdot \\ c_S^i \end{bmatrix} = \bar{\pi}' \begin{bmatrix} \omega_0^i \\ \omega_1^i \\ \cdot \\ \cdot \\ \cdot \\ \omega_S^i \end{bmatrix},$$

$$[1 \ \pi_1 \ \pi_2 \ldots \pi_S] \begin{bmatrix} c_0^i \\ c_1^i \\ \cdot \\ \cdot \\ \cdot \\ c_S^i \end{bmatrix} = [1 \ \pi_1 \ \pi_2 \ldots \pi_S] \begin{bmatrix} \omega_0^i \\ \omega_1^i \\ \cdot \\ \cdot \\ \cdot \\ \omega_S^i \end{bmatrix},$$

$$c_0^i + \sum_{s=1}^{\mathbb{S}} \pi_s c_s^i = \omega_0^i + \sum_{s=1}^{\mathbb{S}} \pi_s \omega_s^i.$$

4.2 First, we know that the market is incomplete because

$$\text{rank}(A) = 2 \leq S = 3.$$

Let's try to find a replicating portfolio

$$\theta = \begin{pmatrix} n \\ m \end{pmatrix}, \qquad n \begin{pmatrix} 1 \\ 1 \\ 1 \end{pmatrix} + \begin{pmatrix} 4 \\ 2 \\ 0 \end{pmatrix} = \begin{pmatrix} 2 \\ 0 \\ 0 \end{pmatrix},$$

$$\begin{aligned} n + 4m &= 2 \\ n + 2m &= 0 \qquad \Rightarrow \text{very obvious there is no } (n, m) = \theta' \\ n &= 0 \end{aligned}$$

$$\text{FTAP} \quad \Rightarrow \quad (q_1 \ q_2) = (\pi_1 \ \pi_2 \ \pi_3) \begin{pmatrix} 1 & 4 \\ 1 & 2 \\ 1 & 0 \end{pmatrix} = \pi A = (1, 2)$$

$$\begin{aligned} &\pi_1 + \pi_2 + \pi_3 = 1 \\ &4\pi_1 + 2\pi_2 \quad = 2 \\ &\left.\begin{aligned} \pi_1 + \pi_2 + \pi_3 &= 1 \\ 2\pi_1 + \pi_2 &= 1 \end{aligned}\right\} \Rightarrow \pi_2 = 1 - 2\pi_1, \quad \pi_3 = \pi_1 \end{aligned}$$

π_s's cannot be negative, $\pi_s \geq 0, s = 1, 2, 3$

$$\pi_3 = \pi_1, \quad \pi_2 = 1 - 2\pi_1$$

$$\pi_1 \in [0, \frac{1}{2}], \text{ otherwise } \pi_2 \text{ becomes negative.}$$

In that case, we can find the bounds for prices (*Note that if we assume strictly monotonic utility functions, then π_s must be strictly positive.*).
The lower bound is found as:

$$\underline{q} = \min_{\pi_1,\pi_2,\pi_3 \geq 0} 2\pi_1 \quad \text{s.t. } \pi_1 + \pi_2 + \pi_3 = 1 \text{ and } 4\pi_1 + 2\pi_2 = 2$$

$$\underline{q} = \min_{\pi_1,\pi_2,\pi_3 \geq 0} 2\pi_1 \quad \text{s.t.}$$

$$\pi_1 \in [0, \frac{1}{2}], \quad \pi_3 = \pi_1, \quad \pi_2 = 1 - 2\pi_1$$

$$\Rightarrow \pi_1 = 0, \ \pi_2 = 1 \text{ and so } \pi_3 = 0 \Rightarrow \underline{q} = 0$$

The upper bound is found as:

$$\bar{q} = \max_{\pi_1,\pi_2,\pi_3 \geq 0} 2\pi_1 \quad \text{s.t. } \pi_1 + \pi_2 + \pi_3 = 1 \text{ and } 4\pi_1 + 2\pi_2 = 2$$

$$\bar{q} = \max_{\pi_1,\pi_2,\pi_3 \geq 0} 2\pi_1 \quad \text{s.t.}$$

$$\pi_1 \in [0, \frac{1}{2}], \quad \pi_3 = \pi_1, \quad \pi_2 = 1 - 2\pi_1$$

$$\pi_1 \uparrow \text{ and } 4\pi_1 + 2\pi_2 = 2 \ \Rightarrow \ \pi_2 \downarrow \text{ so}$$

$$\pi_2^{opt} = 0 \ \Rightarrow \ 4\pi_1 = 2 \text{ so } \pi_1^{opt} = \frac{1}{2} \text{ and } \pi_3^{opt} = \frac{1}{2}$$

$$\Rightarrow \pi_1 = \frac{1}{2} = \pi_3 \Rightarrow \pi_2 = 0 \Rightarrow \bar{q} = 1 \quad \checkmark \quad \text{same solution}$$

By forming portfolios:

$$\begin{array}{ll} \bar{q}(y) = \min_\theta q'\theta & \text{s.t. } A\theta \geq y \\ \underline{q}(y) = \max_\theta q'\theta & \text{s.t. } A\theta \leq y \end{array} \text{ and } y = (2, 0, 0)'$$

$$\min_{\theta_1,\theta_2} \theta_1 + 2\theta_2 \quad \text{s.t.} \begin{array}{l} \theta_1 + 4\theta_2 \geq 2, \\ \theta_1 + 2\theta_2 \geq 0, \\ \theta_1 \qquad\ \ \geq 0. \end{array}$$

The solution will be $(\theta_1, \theta_2) = (0, \frac{1}{2})$.

The lowest payoff that will dominate the payoff of the new asset

$$A\theta^* = \begin{pmatrix} 2 \\ 1 \\ 0 \end{pmatrix} \geq \begin{pmatrix} 2 \\ 0 \\ 0 \end{pmatrix}.$$

The cost of this portfolio $\theta_1^* + 2\theta_2^* = 1$.
$\bar{q}(y) = 1$ upper bound $\Rightarrow$ this means the value of this new asset cannot exceed the price 1 otherwise there is arbitrage.

For the lower bound

$$\max_{\theta_1, \theta_2} \theta_1 + 2\theta_2 \quad \text{s.t.} \quad \begin{array}{l} \theta_1 + 4\theta_2 \leq 2, \\ \theta_1 + 2\theta_2 \leq 0, \\ \theta_1 \qquad\quad \leq 0. \end{array}$$

$\theta_1 = 0, \quad \theta_2 = 0$
$\underline{q}(y) = 0 \rightarrow$ lower bound.

4.3

(a) (i) A strategy that satisfies:

$$\begin{pmatrix} -1 & -2 \\ 1 & 3 \\ 1 & 1 \end{pmatrix} \begin{bmatrix} \theta_1 \\ \theta_2 \end{bmatrix} > 0,$$

cannot be found in this case, since we have three inequalities:

$$\begin{array}{ll} (1)\, \theta_1 + 2 & \theta_2 < 0 \\ (2)\, \theta_1 + 3 & \theta_2 > 0 \\ (3)\, \theta_1 + & \theta_2 > 0. \end{array}$$

Substracting (3) from (2) implies $\theta_2 > 0$ and substracting (3) from (1) implies $\theta_2 < 0$. Thus the asset prices are arbitrage-free.

(ii) A strategy that satisfies:

$$\begin{pmatrix} -1 & -1/2 \\ 1 & 3 \\ 1 & 1 \end{pmatrix} \begin{bmatrix} \theta_1 \\ \theta_2 \end{bmatrix} > 0,$$

can be found in this case. One example of such strategy is: $[\theta_1, \ \theta_2] = [-1, \ 2]$ which yields zero initial investment and 5 payoff in state 1 and 1 payoff in state 2. Prices are not arbitrage-free.

(iii) A strategy that satisfies:

$$\begin{pmatrix} -1 & -1 \\ 1 & 3 \\ 1 & 1 \end{pmatrix} \begin{bmatrix} \theta_1 \\ \theta_2 \end{bmatrix} > 0,$$

can be found in this case. One example of such strategy is: $[\theta_1, \ \theta_2] = [-1, \ +1]$, which yields zero initial investment and payoff 2 in state 1 and payoff 0 in state 2. Prices are arbitrage-free for consumers with weakly monotonic utility functions, but not arbitrage-free for consumers with strictly monotonic utility functions.

(b) (i) Recalling the FTAP pricing formula

$$q = A'\pi,$$

we attain

$$\begin{bmatrix} 1 \\ 2 \end{bmatrix} = \begin{bmatrix} 1 & 1 \\ 3 & 1 \end{bmatrix} \begin{bmatrix} \pi_1 \\ \pi_2 \end{bmatrix},$$

yields $[\pi_1, \ \pi_2] = [1/2, \ 1/2] >> 0$. Hence markets are arbitrage-free.

(ii)

$$\begin{bmatrix} 1 \\ 1/2 \end{bmatrix} = \begin{bmatrix} 1 & 1 \\ 3 & 1 \end{bmatrix} \begin{bmatrix} \pi_1 \\ \pi_2 \end{bmatrix},$$

yields $[\pi_1, \ \pi_2] = [-1, \ 2]$. Hence, markets are not arbitrage-free.

(iii)

$$\begin{bmatrix} 1 \\ 1 \end{bmatrix} = \begin{bmatrix} 1 & 1 \\ 3 & 1 \end{bmatrix} \begin{bmatrix} \pi_1 \\ \pi_2 \end{bmatrix},$$

yields $[\pi_1, \ \pi_2] = [0, \ 1] > 0$. Hence, markets are arbitrage-free only for consumers with weakly monotonic utility functions, but not arbitrage-free for consumers with strictly monotonic utility functions.

(c) (i) The call option payoff on asset 2 with strike 1.5 is

$$\begin{bmatrix} (3 - 1.5)^+ \\ (1 - 1.5)^+ \end{bmatrix} = \begin{bmatrix} 1.5 \\ 0 \end{bmatrix}.$$

(ii)

$$\begin{bmatrix}1.5\\0\end{bmatrix} = m\begin{bmatrix}1\\1\end{bmatrix} + n\begin{bmatrix}3\\1\end{bmatrix}.$$

Yes, it is possible, we can find $(n,\ m) = (-0.75,\ \ 0.75)$ portfolio weights will replicate the call option's payoff.

(iii) The cost of the replicating portfolio will give the price of the new asset:

$$[q_1,\ q_2]\begin{bmatrix}m\\n\end{bmatrix} = [1,\ 2]\begin{bmatrix}m\\n\end{bmatrix} = -0.75 + 2 * 0.75 = 0.75.$$

4.4

(a)

$$\text{rank}(A) = 1 < S = 2 \rightarrow \text{markets are incomplete}$$

Markets are incomplete so we cannot use the (reduced) budget set and solve for state prices.

(b) Financial markets equilibrium

$$\max \ln c_0^i + \frac{1}{2}(\ln c_1^i + \ln c_2^i)$$

$$c_0^i \leq w_0^i - q\theta^i,$$
$$c_1^i \leq w_1^i + A_1\theta^i,$$
$$c_2^i \leq w_2^i + A_2\theta^i,$$
$$\text{where} \quad A = \begin{bmatrix}1\\1\end{bmatrix},$$
$$w^1 = (1, 1, 2),$$
$$w^2 = (1, 2, 1).$$

Investor 1

$$\text{Budget constraint} \;\; c_0^1 + q\theta^1 = w_0^1 = 1,$$
$$c_1^1 = \theta^1 + w_1^1 = \theta^1 + 1,$$
$$c_2^1 = \theta^1 + w_2^1 = \theta^1 + 2,$$

$$\max_{\theta} \ln(1 - q\theta^1) + \frac{1}{2}(\ln(\theta^1 + 1) + \ln(\theta^1 + 2))$$

$$\text{FOC} \quad \frac{\partial U}{\partial \theta} = \frac{-q}{1 - q\theta^1} + \frac{1}{2}\left(\frac{1}{\theta^1 + 1} + \frac{1}{\theta^1 + 2}\right) = 0$$

$$\frac{2q}{1 - q\theta^1} = \frac{1}{\theta^1 + 1} + \frac{1}{\theta^1 + 2}$$

Investor 2

Budget constraint $c_0^2 + q\theta^2 = w_0^2 = 1,$ $\quad\theta^2$: portfolio of the second agent (Investor)

$$c_1^2 = \theta^2 + w_1^2 = \theta^2 + 2,$$

$$c_2^2 = \theta^2 + w_2^2 = \theta^2 + 1,$$

$$\max_{\theta} \ln(1 - q\theta^2) + \frac{1}{2}(\ln(\theta^2 + 2) + \ln(\theta^2 + 1))$$

$$\text{FOC} \quad \frac{\partial U}{\partial \theta} = \frac{-q}{1 - q\theta^2} + \frac{1}{2}\left(\frac{1}{\theta^2 + 2} + \frac{1}{\theta^2 + 1}\right) = 0$$

$$\left.\begin{array}{l} \text{FOC Inv 2} \Rightarrow \frac{2q}{1-q\theta^2} = \frac{1}{\theta^2+2} + \frac{1}{\theta^2+1} \\ \text{FOC Inv 1} \Rightarrow \frac{2q}{1-q\theta^1} = \frac{1}{\theta^1+1} + \frac{1}{\theta^1+2} \end{array}\right\} \text{by symmetry } \theta^1 = \theta^2$$

Denoting without superscript, $\theta^1 = \theta^2 = \theta$, we get

$$\frac{2q}{1 - q\theta} = \frac{(\theta + 2) + (\theta + 1)}{(\theta + 1)(\theta + 2)}$$

$$2q(\theta + 1)(\theta + 2) = 2\theta + 3 - q\theta(2\theta + 3)$$

$$2q(\theta + 1)(\theta + 2) + q\theta(2\theta + 3) = 2\theta + 3$$

$$q = \frac{2\theta + 3}{2(\theta + 1)(\theta + 2) + \theta(2\theta + 3)}$$

We take zero net supply of assets which implies

$$\theta^1 + \theta^2 = 0 \Rightarrow \text{implies } \theta^1 = \theta^2 = 0,$$

i.e. there is no trade. Then the bond price is

$$q = \frac{3}{4}.$$

We know that any state price that will satisfy $\pi_1 + \pi_2 = \frac{3}{4}$ will lead to no-arbitrage prices.

(c) To achieve complete markets we need a payoff vector that is not colinear to the bond payoff, thus, we should not be able to write the payoff of the new asset as a linear function of the old asset(s)

$$\nexists a \in \mathbb{R} \quad \text{s.t. } A^2 = aA^1 = a\begin{bmatrix}1\\1\end{bmatrix} \quad \text{where } A^1, A^2 \in \mathbb{R}^2$$

Then any payoff that pays off different in each state will complete the market.

$$\forall x, y \in \mathbb{R}_+ \quad \text{s.t. } x \neq y \quad \text{then } A = \begin{bmatrix}1 & x\\1 & y\end{bmatrix} \Rightarrow \text{rank}(A) = 2 = S$$

(d)

$$\pi_1 + \pi_2 = \frac{3}{4}$$

and it is given that $x = 2$ and $y = \frac{1}{2}$. Any π_1, π_2 satisfying $\pi_1 + \pi_2 = \frac{3}{4}$ will provide a no-arbitrage price, $2\pi_1 + \frac{1}{2}\pi_2$ so we need to solve for valuation bounds.

Upper bound

$$\max_{\pi_1,\pi_2} 2\pi_1 + \frac{\pi_2}{2} \quad \text{s.t. } \pi_1 + \pi_2 = \frac{3}{4}$$

$$\max_{\pi_2} 2(\underbrace{\frac{3}{4} - \pi_2}_{=\pi_1}) + \frac{\pi_2}{2} = \frac{3}{2} - 2\pi_2 + \frac{\pi_2}{2} = \frac{3}{2} - \frac{3}{2}\pi_2$$

$$\pi_2^{opt} = 0 \quad \pi_1 = \frac{3}{4}$$

$$\bar{q} = 2\frac{3}{4} = \frac{3}{2}$$

Lower bound

$$\min_{\pi_1,\pi_2} 2\pi_1 + \frac{\pi_2}{2} \quad \text{s.t. } \pi_1 + \pi_2 = \frac{3}{4}$$

$$\Rightarrow \min \frac{3}{2} - \frac{3}{2}\pi_2$$

$$\pi_1 = \frac{3}{4} - \pi_2$$

$\Rightarrow$ as No-Arbitrage require π_1 or π_2 cannot be negative then π_2 is bounded above from $\frac{3}{4}$, $\pi_2 \in [0, \frac{3}{4}]$, then $\pi_2^{opt} = \frac{3}{4}$, $\pi_1^{opt} = 0$. Hence, the lower bound

$$\Rightarrow \underline{q} = \frac{3}{4}\frac{1}{3} = \frac{3}{8}.$$

(e) From the previous class, we solved the equilibrium problem for the same setting with complete markets. We found

$$c_0^1 = \frac{1+\pi_1+2\pi_2}{2},\ c_1^1 = \frac{c_0^1}{2\pi_1},\ c_2^1 = \frac{c_0^1}{2\pi_2},$$
$$c_0^2 = \frac{1+2\pi_1+\pi_2}{2},\ c_1^1 = \frac{c_0^2}{2\pi_1},\ c_2^2 = \frac{c_0^2}{2\pi_2}.$$

Market clearing conditions

$$c_0^1 + c_0^2 = w_0^1 + w_0^2 = 2 \Rightarrow 3\pi_1 + 3\pi_2 = 2$$
$$c_1^1 + c_1^2 = w_1^1 + w_1^2 = 3 \Rightarrow 9\pi_1 - 3\pi_2 = 2$$
$$c_2^1 + c_2^2 = w_2^1 + w_2^2 = 3 \Rightarrow 9\pi_2 - \pi_1 = 2$$
$$\Rightarrow \pi_1 = \pi_2 = \frac{1}{3}$$

(f)

$$q = \pi_1 + \pi_2 = [\pi_1\ \pi_2] \underbrace{\begin{bmatrix} 1 \\ 1 \end{bmatrix}}_{\substack{\text{bond} \\ \text{payoff}}} = \frac{2}{3}$$

What happened to the bond price if switching from incomplete to complete markets?

$$\underbrace{q^I = \frac{3}{4} > q^C = \frac{2}{3}}_{decreased!}$$

(g) Complete market price of the asset with payoff $\begin{pmatrix} x \\ y \end{pmatrix}$

$$q^C = [\pi_1\ \pi_2] \begin{bmatrix} x \\ y \end{bmatrix} = \frac{x+y}{3}$$
$$= \frac{1}{3}2 + \frac{1}{3}\frac{1}{2} = \frac{5}{6},$$

$$\underline{q} = \frac{3}{8},$$

$$\bar{q} = \frac{3}{2}.$$

Hence, we observe the new equilibrium price is within the valuation bounds

$$\underline{q} < q^C < \bar{q}.$$

(h) When there was just a risk-free asset, we show that there was no trade $\theta^1 = \theta^2 = 0$ then $c^1 = w^1, c^2 = w^2$, they just consumed their endowment. When there is risky asset as well, we now have complete markets, unique prices (state) for all assets (new). What about the portfolios? As in the previous class, we have the allocations

$$c_0^1 = \frac{1 + \pi_1 + 2\pi_2}{2} = 1 = w_0^1,$$

$$c_1^1 = \frac{c_0^1}{2\pi_1} = \frac{1}{2\frac{1}{3}} = \frac{3}{2} > w_1^1 = 1,$$

$$c_2^1 = \frac{c_0^1}{2\pi_2} = \frac{1}{2\frac{1}{3}} = \frac{3}{2} < w_2^1 = 2.$$

They observed the opportunity and started trading. Now we have nonzero optimal allocations. Obviously it costs zero, it is positive in state one and it is increasing in consumption (or "increases the consumption") while state 2 is decreasing in consumption. Risk averse investors prefer *smooth* consumption. Reduce the volatility of consumption! Increase the overall utility. Here they traded because they have negatively correlated endowments.

4.5

(a) We can easily see that for $m = 1, n = -1$,

$$n \begin{bmatrix} 1 \\ 1 \\ 1 \end{bmatrix} + m \begin{bmatrix} 3 \\ 1 \\ 1 \end{bmatrix} = \begin{bmatrix} 2 \\ 0 \\ 0 \end{bmatrix}.$$

Thus, the call is redundant and the market remains incomplete. We can also see this from the payoff matrix, which still is not of full rank.

$$\det\left(\begin{bmatrix} 1 & 3 & 2 \\ 1 & 1 & 0 \\ 1 & 0 & 0 \end{bmatrix}\right) = 0$$

(Note that A square matrix A_{nxn} is nonsingular ($det(A) \neq 0$) only if its rank is equal to n.)

We can price the call via replication and thus know that

$$\text{Price Call} = \text{Price Stock} - \text{Price Bond}$$

(b) The second option is still redundant to the first two assets and obviously also to the third asset. Again, for $m = \frac{1}{2}, n = \frac{-1}{2}$ we can write,

$$n\begin{bmatrix}1\\1\\1\end{bmatrix} + m\begin{bmatrix}3\\1\\1\end{bmatrix} = \begin{bmatrix}1\\0\\0\end{bmatrix}.$$

(c) Looking at the same setup as before,

$$n\begin{bmatrix}1\\1\\1\end{bmatrix} + m\begin{bmatrix}3\\1\\1\end{bmatrix} = \begin{bmatrix}0\\0\\1\end{bmatrix},$$

there is a contradiction in the bottom two rows, which implies $n + m = 0$ and $n + m = 1$. Thus, the put is not redundant and the market becomes complete. The payoff matrix including the put is now of rank 3. It is impossible to price the put with the existing assets.

4.6

(a)

$$Rank(A) = S = 2 = K \text{ number of assets}$$

Then the markets are complete.

(b) Investor 1's problem

$$\max_{c_0,c_1,c_2} \ln(c_0^1) + \frac{1}{2}\ln(c_1^1) + \frac{1}{2}\ln(c_2^1)$$

s.t.

$$c_0^1 = 1 - q_1\theta_1^1 - q_2\theta_2^1,$$

$$c_1^1 = 1 + 1\theta_1^1 + 0.5\theta_2^1,$$

$$c_2^1 = 2 + 1\theta_1^1 + 2\theta_2^1.$$

and Investor 2's problem

$$\max_{c_0,c_1,c_2} \ln(c_0^2) + \frac{1}{2}\ln(c_1^2) + \frac{1}{2}\ln(c_2^2)$$

s.t.

$$c_0^2 = 1 - q_1\theta_1^2 - q_2\theta_2^2,$$
$$c_1^2 = 2 + 1\theta_1^2 + 0.5\theta_2^2,$$
$$c_2^2 = 1 + 1\theta_1^2 + 2\theta_2^2.$$

and markets clear

$$\sum_{i=1}^{2}\theta_1^i = 0, \ \sum_{i=1}^{2}\theta_2^i = 0, \text{ assets are in zero net supply.}$$

$$\sum_{i=1}^{2} c_0^i = \sum_{i=1}^{2}\omega_0^i, \ \sum_{i=1}^{2} c_1^i = \sum_{i=1}^{2}\omega_1^i, \ \sum_{i=1}^{2} c_2^i = \sum_{i=1}^{2}\omega_2^i.$$

(c) Both agents' $i = 1, 2$ decision problem in Arrow Debreu economy can be expressed as follows:

$$\max_{c_0,c_1,c_2} \ln(c_0^i) + \frac{1}{2}\ln(c_1^i) + \frac{1}{2}\ln(c_2^i)$$

s.t.

$$(c_0^i - \omega_0^i) + \pi_1(c_1^i - \omega_1^i) + \pi_2(c_2^i - \omega_2^i) \leq 0$$

More specifically, consumer 1's problem is:

$$\max_{c_0,c_1,c_2} \ln(c_0^i) + \frac{1}{2}\ln(c_1^i) + \frac{1}{2}\ln(c_2^i)$$

s.t.

$$(c_0^i - 1) + \pi_1(c_1^i - 1) + \pi_2(c_2^i - 2) \leq 0$$

and consumer 2's problem:

$$\max_{c_0,c_1,c_2} \ln(c_0^i) + \frac{1}{2}\ln(c_1^i) + \frac{1}{2}\ln(c_2^i)$$

s.t.

$$(c_0^i - 1) + \pi_1(c_1^i - 2) + \pi_2(c_2^i - 1) \leq 0$$

and markets clear:

$$\sum_{i=1}^{2} c_0^i = \sum_{i=1}^{2} \omega_0^i, \ \sum_{i=1}^{2} c_1^i = \sum_{i=1}^{2} \omega_1^i, \ \sum_{i=1}^{2} c_2^i = \sum_{i=1}^{2} \omega_2^i.$$

(d) We start with the first order conditions of consumer 1's decision problem in Arrow Debreu economy after writing the lagrange function

$$\mathcal{L} = \ln(c_0^i) + \frac{1}{2}\ln(c_1^i) + \frac{1}{2}\ln(c_2^i) - \kappa((c_0^i - \omega_0^i) + \sum_{s=1}^{S=2} \pi_s(c_s^i - \omega_s^i))$$

$$\frac{\partial \mathcal{L}}{\partial c_0^i} = \frac{1}{c_0^i} - \kappa = 0$$

$$\frac{\partial \mathcal{L}}{\partial c_1^i} = \frac{1}{2c_1^i} - \kappa\pi_1 = 0$$

$$\frac{\partial \mathcal{L}}{\partial c_2^i} = \frac{1}{2c_2^i} - \kappa\pi_2 = 0$$

After some arrangements, we get:

$$\kappa = \frac{1}{c_0^i},$$

$$c_1^i = \frac{c_0^i}{2\pi_1},$$

$$c_2^i = \frac{c_0^i}{2\pi_2}.$$

Substituting all in to the budget constraint of the consumer 1 and consumer 1' endowments will yield:

$$c_0^1 = \frac{1 + \pi_1 + 2\pi_2}{2}.$$

In the same way, Substituting all for $i = 2$ in to the budget constraint of the consumer 2 and consumer 2' endowments will yield:

$$c_0^2 = \frac{1 + 2\pi_1 + \pi_2}{2}.$$

Knowing optimal consumption at time 0, we can express all optimal consumption levels in terms of state prices:

$$c_1^1 = \frac{1 + \pi_1 + 2\pi_2}{4\pi_1}, \quad c_2^1 = \frac{1 + \pi_1 + 2\pi_2,}{4\pi_1}$$

$$c_1^2 = \frac{1 + 2\pi_1 + \pi_2}{4\pi_1}, \quad c_2^2 = \frac{1 + 2\pi_1 + \pi_2}{4\pi_1}$$

Markets clear:

$$\sum_{i=1}^{2} c_0^i = 2, \; \sum_{i=1}^{2} c_1^i = 3, \; \sum_{i=1}^{2} c_2^i = 3.$$

By using the market clearing condition, we achieve two equations for state prices:

$$3\pi_1 + 3\pi_2 = 2,$$

$$9\pi_1 - 3\pi_2 = 2,$$

$$\pi_1 = \pi_2 = \frac{1}{3}.$$

(e) The prices of the financial assets:

$$[q_1 \; q_2] = [\pi_1 \; \pi_2] \begin{bmatrix} 1 & 1/2 \\ 1 & 2 \end{bmatrix}$$

$$= [2/3 \; 5/6]$$

(f)

$$c_0^1 = \frac{1 + 1/3 + 2/3}{2} = 1, \; c_1^1 = \frac{c_0^1}{2\pi_1} = \frac{3}{2}, \; c_2^1 = \frac{3}{2},$$

$$c_0^2 = \frac{1 + 2/3 + 1/3}{2} = 1, \; c_1^2 = \frac{c_0^2}{2\pi_2} = \frac{3}{2}, \; c_2^2 = \frac{3}{2}.$$

(g)

$$1 = c_0^1 = 1 - q_1\theta_1^1 - q_2\theta_2^1 = 1 - \frac{2}{3}\theta_1^1 - \frac{5}{6}\theta_2^1$$

$$\frac{3}{2} = c_1^1 = 1 + 1\theta_1^1 + 0.5\theta_2^1$$

$$\frac{3}{2} = c_2^1 = 2 + 1\theta_1^1 + 2\theta_2^1$$

After some steps we find the asset allocations:

$$\theta_1^1 = \frac{5}{6}, \ \theta_2^1 = -\frac{2}{3}.$$

Market clearing gives us the consumer 2's asset allocations:

$$\theta_1^1 = -\frac{5}{6}, \ \theta_2^1 = \frac{2}{3}.$$

(h) No. There is no time preference of any agents in the economy.

(i) Yes. Agents have different endowments in the different states, as they have exactly the same utility functions, so the want to consume optimally the same levels, thus, they trade in such a way that they smooth their consumption equally across states ending up with the exact same consumption allocation.

(j) (i) We need to form a portfolio that will give exactly the same payoff with the option payoff. Then, the price of this portfolio must be the same as the option's price by no-arbitrage principle. We denote the option as the third security in the market and the payoff vector as A^3

$$A^3 = \begin{bmatrix} max(0, \frac{1}{2} - K) \\ max(0, 2 - K) \end{bmatrix} = \begin{bmatrix} A_1^3 \\ A_2^3 \end{bmatrix} = \begin{bmatrix} 0 \\ 0.9 \end{bmatrix}$$

$$A_1^3 = nA_1^1 + mA_1^2$$

$$A_2^3 = nA_2^1 + mA_2^2$$

$$\Downarrow$$

$$q^3 = nq^1 + mq^2$$

$$0 = n + \frac{1}{2}m$$

$$0.9 = n + 2m$$

$$\Downarrow$$

$$n = -\frac{1}{2}m$$

$$0.9 = -\frac{1}{2}m + 2m \ \Rightarrow \ m = \frac{3}{5}$$

$$\Downarrow$$

$$n = -\frac{3}{10}$$

$$q^3 = -\frac{3}{10}q^1 + \frac{3}{5}q^2 = -\frac{3}{10}\frac{2}{3} + \frac{3}{5}\frac{5}{6} = \frac{3}{10}$$

(ii) Recalling the state prices asset pricing rule and equilibrium values of the state prices, we get the price of the third asset

$$q^3 = \pi_1 A_1^3 + \pi_2 A_2^3 = \frac{1}{3}0 + \frac{1}{3}0.9 = \frac{3}{10}$$

(iii) Recall the risk neutral probabilities:

$$\pi_s^* = \frac{\pi_s}{\pi_1 + \pi_2} = \frac{1/3}{2/3} = \frac{1}{2}, \quad \forall s = 1, 2.$$

$$R_f = \frac{3}{2}$$

$$q^3 = \frac{1}{R_f}\mathbb{E}^{\pi^*}\left[A^3\right]$$

$$= \frac{2}{3}(\frac{1}{2}0 + \frac{1}{2}0.9) = \frac{3}{10}$$

4.7

(a) Let $s \geq 0$ be the amount of money the consumer saves. The maximization problem is then:

$$\max_{\{c_0, c_1\} \geq 0} u(c_0, c_1) \quad \text{s.t. } c_0 = w_0 - s, \; c_1 = w_1 + s, \; s \geq 0.$$

With $u(c_0, c_1) = c_0 c_1 + c_1$ this can be rewritten into a maximization problem in s:

$$\max_{s \in [0, w_0]} (w_0 - s)(w_1 + s) + (w_1 + s).$$

(b) The first order condition (FOC) is:

$$\frac{\partial U}{\partial s} = -(w_1 + s) + (w_0 - s) + 1$$

$$= w_0 - w_1 + 1 - 2s$$

$$= -14 - 2s < 0 \quad \text{for } s \geq 0.$$

For any positive saving s, the consumer wants to save less. Therefore, he does not save, i.e., $s = 0$. Consequently, $c_0 = w_0 = 5$ and $c_1 = w_1 = 20$.

(c) The maximization problem becomes:

$$\max_{\{c_0, c_1\} \geq 0} u(c_0, c_1) \quad \text{s.t. } c_0 = w_0 - s, \; c_1 = w_1 + (1 + r_0)s.$$

With $u(c_0, c_1) = c_0c_1 + c_1$ and by neglecting the positivity constraints of c_0 and c_1, this can be rewritten into a maximization problem in s:

$$\max_s \ (w_0 - s)(w_1 + (1 + r_0)s) + (w_1 + (1 + r_0)s).$$

The FOC is then:

$$\frac{\partial U}{\partial s} = -(w_1 + (1 + r_0)s) + (w_0 - s)(1 + r_0) + (1 + r_0) = 0,$$

leading to

$$s = \frac{(w_0 + 1)(1 + r_0) - w_1}{2(1 + r_0)} = -6.524.$$

The optimal consumption plan is then: $c_0 = w_0 - s = 11.524$ and $c_1 = w_1 + (1 + r_0)s = 13.150$.

(d) Since we have no savings in (b), s must be set to zero:

$$0 = s = \frac{(w_0 + 1)(1 + r_0) - w_1}{2(1 + r_0)},$$

therefore, the corresponding interest rate is $r_0 = \frac{w_1}{w_0+1} - 1 = \frac{7}{3}$.

4.8 The investor maximizes:

$$\max_{\{c_{high}, c_{low}\} \geq 0} \log(c_{high}) + \log(c_{low}), \quad \text{s.t.} \begin{cases} w = p_{Shell}\theta_{Shell} + p_{GM}\theta_{GM}, \\ c_{high} = 5\theta_{Shell} + 1\theta_{GM}, \\ c_{low} = 2\theta_{Shell} + 4\theta_{GM}. \end{cases}$$

Neglecting the positivity constraints, we have:

$$\max_{\theta_{Shell}, \theta_{GM}} \log(5\theta_{Shell} + \theta_{GM}) + \log(2\theta_{Shell} + 4\theta_{GM}), \quad \text{s.t. } \theta_{GM} = \frac{w - p_{Shell}\theta_{Shell}}{p_{GM}},$$

or equivalently,

$$\max_{\theta_{Shell}} \log\left(5\theta_{Shell} + \frac{w - p_{Shell}\theta_{Shell}}{p_{GM}}\right) + \log\left(2\theta_{Shell} + 4\frac{w - p_{Shell}\theta_{Shell}}{p_{GM}}\right).$$

Plugging in the figures leads to:

$$\max_{\theta_{Shell}} \log\left(4\theta_{Shell} + 100\right) + \log\left(400 - 2\theta_{Shell}\right).$$

The first order condition is:

$$\frac{\partial U}{\partial \theta_{Shell}} = \frac{4}{4\theta_{Shell} + 100} + \frac{-2}{400 - 2\theta_{Shell}} = 0,$$

which results in $\theta_{Shell} = 87.5$ and $\theta_{GM} = \frac{w - p_{Shell}\theta_{Shell}}{p_{GM}} = 12.5$. The optimal consumption plan is then $c_{high} = 5\theta_{Shell} + \theta_{GM} = 450$ and $c_{low} = 2\theta_{Shell} + 4\theta_{GM} = 225$ (the positivity constraints are satisfied).

4.9

(a) The decision problem of the representative agent is given by:

$$\max_{\{c_0, c_1\} \geq 0} U(c) = u(c_0) + \delta u(c_1), \quad \text{s.t.} \begin{cases} c_0 = w_0 - s, \\ c_1 = w_1 + (1+r)s. \end{cases}$$

(b) Write the first order condition (neglecting the positivity constraints) for the optimization problem in (a):

$$\frac{\partial U}{\partial s} = -u'(c_0) + \delta(1+r)u'(c_1) = 0.$$

Plugging $u'(c) = 1/c$, we have:

$$c_1 = \delta(1+r)c_0,$$

or equivalently,

$$w_1 + (1+r)s = \delta(1+r)(w_0 - s).$$

Solving it with respect to s, we obtain that

$$s = \frac{\delta(1+r)w_0 - w_1}{(1+\delta)(1+r)},$$

and the nonnegativity constraints for consumption are satisfied:

$$c_0 = w_0 - s = \frac{1}{(1+\delta)(1+r)}[(1+r)w_0 + w_1],$$

$$c_1 = w_1 + (1+r)s = \frac{\delta}{(1+\delta)}[(1+r)w_0 + w_1].$$

The representative agent would save ($s > 0$) if and only if

$$\delta(1+r)w_0 > w_1,$$

or in terms of the income growth:

$$\delta(1+r) > g.$$

Here, $\delta(1+r)$ represents the time discounted riskless return per unit saving. The model suggests that for the agent to save, the interest rate must supply the time discounted return that is higher than the growth rate of agent's income.

4.10 First, we note that the payoff matrix A has a full column rank equal to the number of states. Thus, the financial market is complete. Second, we check whether there is an arbitrage in the market. By the FTAP we need to verify whether there exist state prices π_1 and π_2, such that

$$(100, 100) = (\pi_1, \pi_2) \begin{pmatrix} 100 & 50 \\ 100 & 200 \end{pmatrix}.$$

The unique solution is $\pi_1 = 2/3$ and $\pi_2 = 1/3$, therefore, the market is arbitrage-free.

In a complete arbitrage-free market, a price of any security is uniquely determined.

To price a call option we can use the replicating strategy approach as an example. Let $C = \max(S - K, 0)$ be a call option on the underlying risky asset S (asset 2) with a strike K. Since the market is complete, there exists a corresponding hedging portfolio $(\theta_1, \theta_2)'$:

$$\begin{aligned} &\text{state 1: } \max(50 - K, 0) &&= 100 \cdot \theta_1 + 50 \cdot \theta_2, \\ &\text{state 2: } \max(200 - K, 0) &&= 100 \cdot \theta_1 + 200 \cdot \theta_2. \end{aligned}$$

Solving the above with respect to θ_1, θ_2 gives:

$$\begin{aligned} \theta_1 &= \frac{1}{300}(4\max(50 - K, 0) - \max(200 - K, 0)), \\ \theta_2 &= \frac{1}{150}(\max(200 - K, 0) - \max(50 - K, 0)). \end{aligned}$$

The option price is then given by:

$$q_C = q_1\theta_1 + q_2\theta_2 = \frac{2}{3}\max(50 - K, 0) + \frac{1}{3}\max(200 - K, 0).$$

To price a put option we then use the state price approach. In a complete non-arbitrage market unique state prices π exist. The price of a security is the sum of its payoffs weighted by corresponding state prices. Denote by $P = \max(K - S, 0)$ a put option on the underlying risky asset S (asset 2) with a strike K. Then its price is given by

$$
\begin{aligned}
q_P &= \mathbb{E}_\pi \max(K - S, 0) \\
&= \pi_1 \cdot \max(K - 50, 0) + \pi_2 \cdot \max(K - 200, 0) \\
&= \frac{2}{3} \max(K - 50, 0) + \frac{1}{3} \max(K - 200, 0).
\end{aligned}
$$

4.11

(a) The financial market is incomplete. Indeed, the number of states $S = 3$ is greater than the number of assets $K = 2$, therefore, there exist consumption streams that cannot be replicated by these assets.

In particular, the security with payoffs $(1, 0, 0)'$ cannot be hedged by assets 1 (risk-free asset) and 2 (risky asset). If it were possible, then there would exist a hedging portfolio $\theta = (\theta_1, \theta_2)'$, which generated the same payoffs in all states as the security:

$$
\begin{aligned}
&(1) 1 \cdot \theta_1 + 1 \cdot \theta_2 = 1, \\
&(2) 1 \cdot \theta_1 + 1.5 \cdot \theta_2 = 0, \\
&(3) 1 \cdot \theta_1 + 0.5 \cdot \theta_2 = 0.
\end{aligned}
$$

These equations have no solution, hence the security cannot be hedged. To see this, subtract (3) from (2) to find $\theta_2 = 0$. From (1) it then follows that $\theta_1 = 1$. But then (2) and (3) cannot be solved.

(b) Since the security cannot be hedged by the two assets, there is no unique price of the security. Its upper and lower price bounds can be found, e.g., via dominating and dominated portfolios:

$$
\overline{q} = \min_\theta \; q'\theta \quad \text{s.t.} \begin{cases} 1 \cdot \theta_1 + 1 \cdot \theta_2 \geq 1, \\ 1 \cdot \theta_1 + 1.5 \cdot \theta_2 \geq 0, \\ 1 \cdot \theta_1 + 0.5 \cdot \theta_2 \geq 0, \end{cases} \tag{4.1}
$$

$$
\underline{q} = \max_\theta \; q'\theta \quad \text{s.t.} \begin{cases} 1 \cdot \theta_1 + 1 \cdot \theta_2 \leq 1, \\ 1 \cdot \theta_1 + 1.5 \cdot \theta_2 \leq 0, \\ 1 \cdot \theta_1 + 0.5 \cdot \theta_2 \leq 0. \end{cases} \tag{4.2}
$$

Equation (4.1) states that the upper bound for the security price equals the value of the least expensive dominating portfolio, while (4.2) implies that the low bound for the security price equals the value of the most expensive dominated portfolio. It is straightforward to show that the upper bound for the security price

$\overline{q}$ is equal 1 and can be achieved, e.g., on a dominating portfolio $\overline{\theta} = (0.5, 0.5)'$, see Fig. 4.1.

Similarly, the lower bound for the security price is $\underline{q} = 0$ and can be achieved, e.g., on a dominated portfolio $\underline{\theta}_1 = \underline{\theta}_2 = 0$, see Fig. 4.2.

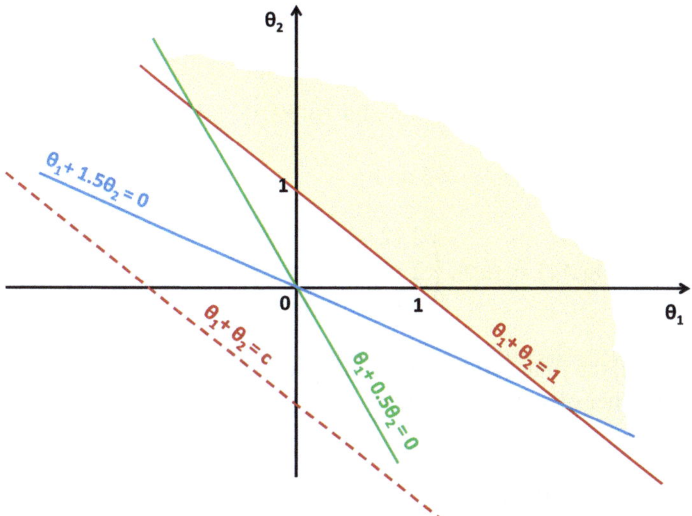

Fig. 4.1 Shaded area corresponds to dominating portfolios. The price of portfolio θ is $\theta_1 + \theta_2$. As can be seen from the graph, the price of the less expensive dominating portfolio is 1

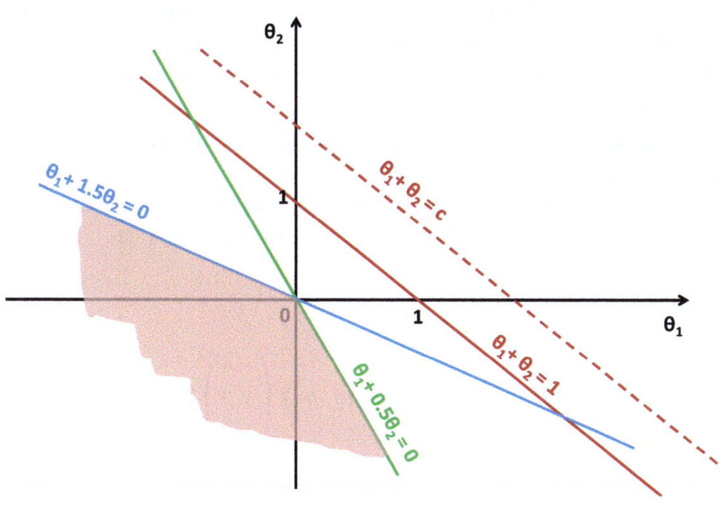

Fig. 4.2 Shaded area corresponds to dominated portfolios. The price of portfolio θ is $\theta_1 + \theta_2$. As can be seen from the graph, the price of the most expensive dominated portfolio is 0

4.12

(a) The financial market is complete, if all second period consumption streams can be achieved by asset trade. Mathematically it means that the number of states S does not exceed the number of assets that have linearly independent payoffs, or equivalently, the number of states S does not exceed the column rank of the payoff matrix A. Recall, that the column rank of the matrix does not change with linear transformations of its columns.

1. The number of states $S = 2$ and $\text{rank}(A_1) = 2$, therefore, the financial market is complete.
2. The number of states $S = 3$ and $\text{rank}(A_2) = 2$, therefore, the financial market is incomplete.
3. The number of states $S = 3$ and

$$\text{rank}(A_3) = \text{rank}\begin{pmatrix} 2 & (2-2\cdot 2) & 2 \\ 2 & (0-2\cdot 1) & 1 \\ 0 & (2-2\cdot 1) & 1 \end{pmatrix} = \text{rank}\begin{pmatrix} 2 & -2 & 2 \\ 2 & -2 & 1 \\ 0 & 0 & 1 \end{pmatrix} = 2,$$

therefore, the financial market is incomplete.

4. The number of states $S = 3$ and

$$\text{rank}(A_4) = \text{rank}\begin{pmatrix} (2-2\cdot 4) & (1-4\cdot 4) & 4 \\ (2-2\cdot 1) & (2-4\cdot 1) & 1 \\ (2-2\cdot 1) & (4-4\cdot 1) & 1 \end{pmatrix} = \text{rank}\begin{pmatrix} -6 & -15 & 4 \\ 0 & -2 & 1 \\ 0 & 0 & 1 \end{pmatrix} = 3,$$

therefore, the financial market is complete.

(b) Recall writing mistake of the Fundamental Theorem of Asset Prices for qualitative different utility functions:

(i) There is no arbitrage for strictly monotonous utilities if and only if there exist state prices $\pi >> 0$, such that $q = A^T\pi$,

(ii) There is no arbitrage for weakly monotonous utilities if and only if there exist state prices $\pi > 0$, such that $q = A^T\pi$,

(iii) There is no arbitrage for mean-variance utilities if and only if there exist state prices $\pi \neq 0$, such that $\sum_{s=1}^{S} \pi_s > 0$ and $q = A^T\pi$.

1. For the state prices,

$$(1, 2) = (\pi_1, \pi_2, \pi_3)\begin{pmatrix} 1 & 4 \\ 1 & 2 \\ 1 & 0 \end{pmatrix},$$

or equivalently, $\pi_2 = 1-2\pi_1, \pi_3 = \pi_1, \pi_1 \in \mathbb{R}$. Assigning, e.g., $\pi_1 = 0.25 > 0$, we get $\pi_2 = 0.5 > 0$ and $\pi_3 = 0.25 > 0$. Therefore, there is no arbitrage opportunity in the market neither for strictly, weakly monotonous nor mean-variance utilities.

2. For the state prices,

$$(1, 1, 1) = (\pi_1, \pi_2, \pi_3) \begin{pmatrix} 2 & 1 & 4 \\ 2 & 2 & 1 \\ 2 & 4 & 1 \end{pmatrix}.$$

The unique solution is $\pi_1 = \frac{1}{6} > 0$, $\pi_2 = \frac{1}{4} > 0$ and $\pi_3 = \frac{1}{12} > 0$. Therefore, there is no arbitrage opportunity in the market for neither strictly, weakly monotonous nor mean-variance utilities.

4.13 Denote by $k = 1$ the risk-free asset and $k = 2$ the risky asset. Then the returns matrix is

$$R = (R_s^k) = \begin{pmatrix} 2 & 10 \\ 2 & 5 \\ 2 & -5 \end{pmatrix}.$$

(a) There are three states and only two assets, therefore the financial market is incomplete.

Let R^λ be the return vector generated by an investment strategy $\lambda = (\lambda_1, \lambda_2)'$, $\lambda_1 + \lambda_2 = 1$. Then

$$R^\lambda = \begin{pmatrix} R_1^\lambda \\ R_2^\lambda \\ R_3^\lambda \end{pmatrix} = R\lambda. \quad (4.3)$$

Since the full rank of the returns matrix R is equal 2, the matrix R has only two linearly independent rows. In particular, its row R_3 can be obtained as a linear combination of the other two rows R_1 and R_2:

$$R_3 = -2R_1 + 3R_2. \quad (4.4)$$

Hence, as follows from (4.3), for any return vector R^λ that can be replicated by assets 1 and 2, the same equality should hold for its components, i.e.,

$$R_3^\lambda = -2R_1^\lambda + 3R_2^\lambda. \quad (4.5)$$

This implies that any return vector whose components do not satisfy (4.5) cannot be hedged by the existing assets. In particular, the return vector $r^{nohedge} = (1, 1, 0)^T$ cannot be hedged in the considered market.

(b) Denote by π_1^*, π_2^* and π_3^* the risk adjusted probabilities. Then the expected return μ of the risky asset is equal to $\pi_1^* \cdot 10\% + \pi_2^* \cdot 5\% + \pi_3^* \cdot (-5\%)$. Hence, $\mu = r_f$ is and only if

$$2 = 10\pi_1^* + 5\pi_2^* - 5\pi_3^*. \quad (4.6)$$

Besides, for the probabilities

$$\pi_1^* + \pi_2^* + \pi_3^* = 1. \tag{4.7}$$

Therefore, combining (4.6) and (4.7) we get the set of required risk adjusted probabilities:

$$\pi_1^* \in R, \quad \pi_2^* = \frac{7}{10} - \frac{3}{2}\pi_1, \quad \pi_3^* = \frac{3}{10} + \frac{1}{2}\pi_1.$$

(c) In part (b) we already derived the set of risk adjusted probabilities such that the expected return of the first risky asset is equal to the risk-free rate. Next we narrow this set of risk adjusted probabilities, so that the expected return of the second risky asset is equal to the risk-free rate:

$$\pi_1^* \cdot 1\% + \pi_2^* \cdot 2\% + \pi_3^* \cdot 0\% = 2\%,$$

or equivalently, $\pi_2^* = 1 - \frac{1}{2}\pi_1^*$. Combining it with (b), we obtain that

$$\pi_1^* = -\frac{3}{10}, \quad \pi_2^* = \frac{23}{20}, \quad \pi_3^* = \frac{3}{20}. \tag{4.8}$$

In Exercise 4.3 we have considered no-arbitrage conditions for qualitative different utility functions. To apply them one need to find the set of all possible state prices π, such that $q = A^T\pi$. Recall, that

$$\begin{aligned}
q = A^T\pi \qquad & \Longleftrightarrow q^k = \sum_{s=1}^{S} A_s^k \pi_s,\ k = 0, \ldots, K, \\
& \Longleftrightarrow \begin{cases} 1 = R_f \sum_{s=1}^{S} \pi_s, \\ 1 = \sum_{s=1}^{S} R_s^k \pi_s,\ k = 1, \ldots, K, \end{cases} \\
& \Longleftrightarrow \begin{cases} 1 = \sum_{s=1}^{S} \pi_s^*, \\ R_f = \sum_{s=1}^{S} R_s^k \pi_s^*,\ k = 1, \ldots, K, \end{cases} \\
& \Longleftrightarrow \begin{cases} 1 = \sum_{s=1}^{S} \pi_s^*, \\ r_f = \sum_{s=1}^{S} r_s^k \pi_s^*,\ k = 1, \ldots, K, \end{cases}
\end{aligned}$$

where π_s^* are risk adjusted or risk neutral probabilities, $\pi_s^* = \pi_s R_f$, and R_s^k and r_s^k is the gross and nett return of asset k in state s correspondingly.

In (4.8) we derived the unique risk neutral probabilities π^* for the considered market. The state prices π, $\pi_s = \frac{\pi_s^*}{R_f}$, are therefore also unique and are given by

$$\pi_1 = -\frac{5}{17}, \quad \pi_2 = \frac{115}{102}, \quad \pi_3 = \frac{5}{34}.$$

Since the price $\pi_1 < 0$, while $\pi_1 + \pi_2 + \pi_3 = \frac{1}{1.02} > 0$, there exists an arbitrage opportunity for strictly and weakly monotonous utility functions, however, there is no arbitrage for mean-variance utilities.

4.14 For assets A and B, with two possible states occurring with prob p and $(1-p)$,

$$\begin{aligned} E(A) &= pA_1 + (1-p)A_2 = 0.2 \\ E(B) &= pB_1 + (1-p)B_2 = 0.1 \\ V(A) &= p(A_1 - 0.2)^2 + (1-p)(A_2 - 0.2)^2 = 0.3 \\ V(B) &= p(B_1 - 0.1)^2 + (1-p)(B_2 - 0.1)^2 = 0.2 \\ COV(A, B) &= p(A_1 - 0.2)(B_1 - 0.1) + (1-p)(A_2 - 0.2)(B_2 - 0.1) \end{aligned}$$

Plug in the numbers:

$$\begin{aligned} E(A) &= 0.5A_1 + 0.5A_2 = 0.2 \\ E(B) &= 0.5B_1 + 0.5B_2 = 0.1 \\ V(A) &= 0.5(A_1 - 0.2)^2 + 0.5(A_2 - 0.2)^2 = 0.3 \\ V(B) &= 0.5(B_1 - 0.1)^2 + 0.5(B_2 - 0.1)^2 = 0.2 \end{aligned}$$

From (6) and (7) we express A_1 and B_1 and substitute into (8) and (9):

$$(0.2 - A_2)^2 = 0.3 \quad and \quad (0.1 - B_2)^2 = 0.2$$

Thus, we have to solve (with the determinants shown in brackets):

$$\begin{aligned} A_2^2 - 0.4A_2 - 0.26 = 0 \quad (D_A = 1.2) \\ B_2^2 - 0.2B_2 - 0.19 = 0 \quad (D_B = 0.8) \end{aligned}$$

Each equation has two roots (call it positive and negative), and we receive two pairs of solutions:

$$A_2^+ = 0.747 \quad \text{and} \quad A_1^+ = -0.347; \quad \text{or} \quad A_2^- = -0.347 \quad \text{and} \quad A_1^- = -0.747$$

$$B_2^+ = 0.547 \quad \text{and} \quad B_1^+ = -0.347; \quad \text{or} \quad B_2^- = -0.347 \quad \text{and} \quad B_1^- = 0.547$$

To choose a particular solution, we note that the covariance between the two returns is specified to be positive, in other words, in state s, whenever A is high, B should also be high, and the other way around. Hence, the solution is

either $\begin{pmatrix} -0.347 & -0.347 \\ 0.747 & 0.547 \end{pmatrix}$ or $\begin{pmatrix} 0.747 & 0.547 \\ -0.347 & -0.347 \end{pmatrix}$

That the covariance is also satisfied can be checked easily.

4.15

(a)

$$E(R^1) = \frac{1}{2}(\mu_1 + \rho\sigma_1 + \mu_1 - \rho\sigma_1) = \mu_1$$

$$E(R^2) = \frac{1}{2}(\mu_2 - \sigma_2 + \mu_2 + \sigma_2) = \mu_2$$

$$V(R^1) = \frac{1}{2}(\mu_1 + \rho\sigma_1 - \mu_1)^2 + \frac{1}{2}(\mu_1 - \rho\sigma_1 - \mu_1)^2 = \rho^2\sigma_1^2$$

$$V(R^2) = \sigma_1^2$$

$$\mathrm{COR}(R^1, R^2) = \frac{\mathrm{cov}(R^1, R^2)}{\sqrt{V(R^1)\cdot V(R^2)}} = \frac{1}{2}(-\rho\sigma_1\sigma_2 + (-\rho\sigma_1\sigma_2))/\rho\sigma_1\sigma_2 = -1$$

(b) Let $\mu_1, \mu_2 \geq 1$, i.e. assets have non-negative expected net return. We need to consider four cases in our parameter space.

(A) $\mu_1 - \rho\sigma_1 < 1, \mu_2 - \sigma_2 < 1$; i.e. $\rho > \frac{\mu_1 - 1}{\sigma_1}$

$$\widehat{R} = \begin{pmatrix} N & \mu_2 - \sigma_2 \\ \mu_1 - \rho\sigma_1 & N \end{pmatrix}; \; \widehat{R}^P = max\begin{pmatrix} \frac{N+\mu_2-\sigma_2}{2}, 1 \\ \frac{N+\mu_1-\rho\sigma_1}{2}, 1 \end{pmatrix}$$

Both outcomes are equally likely to occur, none is equal to N.

(B) $\mu_1 - \rho\sigma_1 < 1, \mu_2 - \sigma_2 \geq 1$

$$\widehat{R} = \begin{pmatrix} N & N \\ \mu_1 - \rho\sigma_1 & N \end{pmatrix}; \; \widehat{R}^P = max\begin{pmatrix} N, 1 \\ \frac{N+\mu_1-\rho\sigma_1}{2}, 1 \end{pmatrix}$$

Here the probability of getting return N (given that $N > 1$) is equal to 0.5 (outcome 1).

(C) $\mu_1 - \rho\sigma_1 \geq 1, \mu_2 - \sigma_2 < 1$; i.e. $\rho \leq \frac{\mu_1 - 1}{\sigma_1}$

$$\widehat{R} = \begin{pmatrix} N & \mu_2 - \sigma_2 \\ N & N \end{pmatrix}; \; \widehat{R}^P = max\begin{pmatrix} \frac{N+\mu_2-\sigma_2}{2}, 1 \\ N, 1 \end{pmatrix}$$

Here the probability of getting return N (outcome 2) is equal to 0.5.

(D) $\mu_1 - \rho\sigma_1 \geq 1, \mu_2 - \sigma_2 \geq 1$

$$\widehat{R} = \begin{pmatrix} N & N \\ N & N \end{pmatrix}; \quad \widehat{R}^P = max \begin{pmatrix} N, 1 \\ N, 1 \end{pmatrix}$$

Outcome N always attained.

4.16

(a) The state-space matrix is given by the returns of the assets (k = stocks, bonds) in the states determined by the two factors (f = oil price, growth rate). By assumption, each factor has two realizations which we denote as "high" and "low". Hence, there are four states determining the returns of the assets:
1. growth rate="high", oil price="high",
2. growth rate="high", oil price="low",
3. growth rate="low", oil price="high",
4. growth rate="low", oil price="low".

The state-space matrix is an $S \times K$ matrix, where S denotes the number of states and K is the number of assets. In our example, $S = 4$ and $K = 2$. Given the returns of each asset for each of the two factors, we get that the state-space matrix is

$$R = (R_s^k) = \begin{pmatrix} 1\% & -3\% \\ 5\% & -1\% \\ -2\% & -1\% \\ -3\% & 2\% \end{pmatrix}.$$

(b) To find the factor loadings consider that the state-space matrix is the product of the factor values in each state and the factor loadings, i.e.

$$R = [R_s^k] = (R_s^f)(\beta_k^f)'. \tag{4.9}$$

The factor returns R_s^f cannot be set completely arbitrarily. Indeed, if the returns matrix R has linearly dependent rows, the same relation should hold for the rows of the factor values matrix (R_s^f). In our example the rank of (R_s^k) is equal 2, hence only two of its rows are linearly independent and, e.g., the assets returns in states 3 and 4 may be expressed as linear combinations of the assets returns in states 1 and 2:

$$R_3 = 0.5R_1 - 0.5R_2, \quad R_4 = -0.5R_1 - 0.5R_2. \tag{4.10}$$

Therefore, the same linear equalities (4.10) hold for the rows of the factor values matrix (R^s_f):

$$FV_3 = 0.5FV_1 - 0.5FV_2, \quad FV_4 = -0.5FV_1 - 0.5FV_2. \tag{4.11}$$

Given (R^k_s) as calculated in the previous question, for given factor values summarized in the matrix (R^f_s) as, e.g.,

$$(R^f_s) = \begin{pmatrix} 7 & -3 \\ 7 & -1 \\ 0 & -1 \\ -7 & 2 \end{pmatrix},$$

we get that for (4.9) to hold the factor loadings must be

$$(\beta^f_k) = \begin{pmatrix} 1 & 2 \\ 0 & 1 \end{pmatrix}.$$

(c) The state-space matrix summarizes the asset returns in different states. To determine the joint distribution of the asset returns, we need to consider the returns of each asset across the states and the joint probability of factor combinations generating these states. As an example,

$$P(stocks = -2\%, bonds = -1\%) = P(growth = low, oil = high) = 0.5.$$

Thus, the joint distribution of the asset returns is given by the following table.

Stocks, bonds	−3%	−1%	2%
1%	0.05	0	0
5%	0	0.3	0
−2%	0	0.5	0
−3%	0	0	0.15

4.17

(a) To calculate the state-space matrix we use Eq. (4.9) and get

$$R = \begin{pmatrix} 3\% & -2\% \\ 2\% & -1\% \\ -1\% & 2\% \\ 2\% & 4\% \end{pmatrix} \cdot \begin{pmatrix} 1 & -2 \\ -3 & 1 \end{pmatrix} = \begin{pmatrix} 9\% & -8\% \\ 5\% & -5\% \\ -7\% & 4\% \\ -10\% & 0\% \end{pmatrix},$$

where $k = 1$ stands for stocks and $k = 2$ for bonds.

(b) If we assume that the four states are equally probable, then the joint distribution of asset returns is given by

Stocks, bonds	−8%	−5%	0%	4%
9%	0.25	0	0	0
5%	0	0.25	0	0
−7%	0	0	0	0.25
−10%	0	0	0.25	0

As an example,

$$\begin{aligned} P(stocks = 9\%, bonds = -8\%) &= P(growth = 3\%, inflation = -2\%) \\ &= P(state = 1) = 1/4. \end{aligned}$$

4.18

(a) The mean and the variance of the market and the mean of the investors portfolio are:

$$\mu_M = 2 \qquad \sigma_M^2 = \frac{32}{3} \qquad \mu_I = 6$$

Determine the β_I of the investors portfolio:

$$\beta_I = \frac{\operatorname{cov}(r_I, r_M)}{\sigma_M^2} = \frac{\frac{1}{3}((6-2)(23-6) + (2-2)(-4-6) + (-2-2)(-1-6))}{\frac{32}{3}} = 3$$

If the portfolio is feasible the security market line is satisfied:

$$\mu_i - R_f = \beta_i(\mu_M - R_f)$$

Check that the SML is satisfied

$$6 - 0 = 3(2 - 0)$$

Indeed the SML is satisfied. I.e. the portfolio is feasible.

(b) No, the portfolio is not optimal since the portfolio of the investor and the market is not comonotone (see proposition from the lecture). To see that look at state 2 and 3 of the two portfolios.

(c) Since all states are equal likely the investor is indifferent between the following portfolios:

$$r_I = \begin{pmatrix} 23\% \\ -4\% \\ -1\% \end{pmatrix} \qquad r_{I'} = \begin{pmatrix} 23\% \\ -1\% \\ -4\% \end{pmatrix}$$

r_I is the initial portfolio of the investor and $r_{I'}$ is a portfolio with the same expected utility as the initial portfolio and $r_{I'}$ is comonotone to the market.

$r_{I'}$ is not on the SML. Now we need to add an appropriate constant to $r_{I'}$ such that $r_{I^*} := r_{I'} + \text{constant}$ is on the SML. Note that the β of $r_{I'}$ and r_{I^*} is the same due to the fact that they only differ by a constant.

Determine $\beta_{I'}$:

$$\beta_{I'} = \frac{\operatorname{cov}(r_{I'}, r_M)}{\sigma_M^2}$$
$$= \frac{\frac{1}{3}((6-2)(23-6) + (2-2)(-1-6) + (-2-2)(-4-6))}{\frac{32}{3}} = \frac{27}{8}$$

Since the portfolio I^* has to lie on the SML and $\beta_{I^*} = \beta_{I'}$:

$$\mu_{I^*} - R_f = \beta_{I^*}(\mu_m - R_f) = \beta_{I'}(\mu_m - R_f)$$
$$= \frac{27}{8}(2-0) = \frac{27}{4} = 6.75$$

Since $\mu_{I'} = 6$ and $\mu_{I^*} = 6.75$

$$r_{I^*} = r_{I'} - \mu_{I'} + \mu_{I^*}$$
$$= \begin{pmatrix} 23\% \\ -1\% \\ -4\% \end{pmatrix} - 6\% + 6.75\% = \begin{pmatrix} 23\% \\ -1\% \\ -4\% \end{pmatrix} + 0.75\% = \begin{pmatrix} 23.75\% \\ -0.25\% \\ -3.25\% \end{pmatrix}$$

In every state r_{I^*} pays more than $r_{I'}$ and is therefore preferred to $r_{I'}$ and therefore also to the initial portfolio r_I. Furthermore it is feasible.

4.19

(a) First it needs to be shown that expected utility preferences can be expressed as mean variance preferences in the case of normal distributed returns.

$$\mathbb{E}(u(X)) = \int_{-\infty}^{\infty} u(x) f(x; \mu, \sigma^2) dx$$
$$= \int_{-\infty}^{\infty} u(x) \frac{1}{\sqrt{2\pi\sigma^2}} \exp\left(-\frac{1}{2}\frac{(x-\mu)^2}{\sigma^2}\right) dx$$

Substitute $x = \mu + \sigma z$

$$= \int_{-\infty}^{\infty} u(\mu + \sigma z)\sqrt{2\pi} \exp\left(-\frac{1}{2}z^2\right) dx$$

with $Z = N(0, 1)$ this can be written as

$$\begin{aligned} &= \mathbb{E}\left(u\left(\mu + \sigma Z\right)\right) \\ &=: V(\mu, \sigma) \end{aligned}$$

i.e. $\mathbb{E}(u(X))$ can be written as a function in μ and σ.

(b) If $V(\mu, \sigma)$ is increasing in μ then the derivative of $V(\mu, \sigma)$ to μ must be positive

$$\frac{\partial V(\mu, \sigma)}{\partial \mu} = \frac{\partial \mathbb{E}\left(u\left(\mu + \sigma Z\right)\right)}{\partial \mu} = \mathbb{E}\left(u'\left(\mu + \sigma Z\right)\right)$$

The condition that the utility is increasing in μ is then

$$\mathbb{E}\left(u'\left(\mu + \sigma Z\right)\right) > 0$$

This condition is for example satisfied if u is a strictly increasing function. The condition that the utility decreases with σ is obtained in the same manner

$$\frac{\partial V(\mu, \sigma)}{\partial \sigma} = \frac{\partial \mathbb{E}\left(u\left(\mu + \sigma Z\right)\right)}{\partial \sigma} = \mathbb{E}\left(u'\left(\mu + \sigma Z\right) Z\right) < 0$$

The intuition of this condition is not clear. Concavity of the utility functions helps to satisfy this condition, but it is not enough.

(c) Now it needs to be shown, that also a CPT-utility function can be written as a mean-variance utility function, if the returns are normally distributed:

$$\begin{aligned} \mathrm{CPT}(X) &= \int_{-\infty}^{\infty} v(x) \left(\left. \frac{dw(F(y))}{dy} \right|_{y=x} \right) dx \\ &= \int_{-\infty}^{\infty} v(x) w'(F(x)) f(x) dx \\ &= \int_{-\infty}^{\infty} v(x) w'(F(x)) \frac{1}{\sqrt{2\pi\sigma^2}} \exp\left(-\frac{1}{2}\frac{(x-\mu)^2}{\sigma^2}\right) dx \end{aligned}$$

Substitute $x = \mu + \sigma z$

$$= \int_{-\infty}^{\infty} v(\mu + \sigma z) w'(F(\mu + \sigma z)) \sqrt{2\pi} \exp\left(-\frac{1}{2} z^2\right) dx$$

with $Z = N(0, 1)$ this can be written as

$$\begin{aligned} &= \mathbb{E}\left[v(\mu + \sigma Z) w'(F(\mu + \sigma Z))\right] \\ &=: V(\mu, \sigma) \end{aligned}$$

I.e. in the case of normally distributed returns the CPT-utilities can be expressed as mean-variance-utilities. Note that the conditions under which the utilities are increasing in μ and decreasing in σ are much more complicated in the CPT-case than in the expected utility case.

4.20

(a) The price of the risky asset can be determined from the equilibrium of this economy. Hence, we need to determine the equilibrium allocations and prices. The equilibrium in this economy is defined as follows:

The representative agent maximizes his utility function under budget constraints:

$$\theta = argmax\ U = \frac{1}{2}\ln(c_u) + \frac{1}{2}\ln(c_d)$$

s.t.

$$c_u = \theta_0 q^0 R + \theta_1 D(\mu + \sigma) \tag{1}$$

$$c_d = \theta_0 q^0 R + \theta_1 D(\mu - \sigma) \tag{2}$$

$$\theta_0 q^0 + \theta_1 S_0 = \theta_{A,0} q^0 + \theta_{A,1} S_0 = \theta_{A,1} S_0 \tag{3}$$

where q^0 represents the price of the risky asset, and $\theta_{A,k}$ stands for the initial endowment in asset k. Market clears:

$$\theta = \theta_A$$

Expressing θ_0 from the budget constraint (3) and substituting it in constraints (1) and (2) and finally in the utility function we get:

$$\theta_0 q^0 + \theta_1 S_0 = \theta_{A,1} S_0 \ \rightarrow\ \theta_0 = \frac{S_0}{q^0}\left(\theta_{A,1} - \theta_1\right) \tag{3}$$

$$c_u = S_0\left(\theta_{A,1} - \theta_1\right) R + \theta_1 D(\mu + \sigma) \tag{1}$$

$$c_d = S_0(\theta_{A,1} - \theta_1) R + \theta_1 D(\mu - \sigma) \tag{2}$$

$$U = \frac{1}{2}\ln\left(S_0\left(\theta_{A,1} - \theta_1\right) R + \theta_1 D(\mu + \sigma)\right) + \frac{1}{2}\ln\left(S_0\left(\theta_{A,1} - \theta_1\right) R + \theta_1 D(\mu - \sigma)\right)$$

Therefore, the first order condition of this problem is:

$$\frac{\partial U}{\partial \theta_1} = \frac{1}{2}\frac{-S_0 R + D(\mu + \sigma)}{S_0(\theta_{A,1} - \theta_1)R + \theta_1 D(\mu + \sigma)} + \frac{1}{2}\frac{-S_0 R + D(\mu - \sigma)}{S_0(\theta_{A,1} - \theta_1)R + \theta_1 D(\mu - \sigma)} = 0$$

Substituting the market clearing condition into the F.O.C. we get:

$$\theta = \theta_A = \begin{bmatrix} 0 \\ 1 \end{bmatrix}$$

$$\frac{\partial U}{\partial \theta_1} = \frac{1}{2}\frac{-S_0 R + D(\mu+\sigma)}{\theta_1 D(\mu+\sigma)} + \frac{1}{2}\frac{-S_0 R + D(\mu-\sigma)}{\theta_1 D(\mu-\sigma)} = 0$$

$$\frac{-S_0 R + D(\mu+\sigma)}{(\mu+\sigma)} = \frac{S_0 R + D(\mu-\sigma)}{(\mu-\sigma)}$$

$$-S_0 R(\mu-\sigma) + D(\mu^2-\sigma^2) = S_0 R(\mu+\sigma) - D(\mu^2-\sigma^2)$$

$$S_0 = \frac{D(\mu^2-\sigma^2)}{R\mu}$$

(b) From the obtained formula for S_0 we can see that it decreases in σ. Namely:

$$\frac{\partial S_0}{\partial \sigma} = \frac{-2D\sigma}{R\mu}$$

Since σ can be seen as the volatility of the dividend, it is always a positive value. Assuming that$(D \cdot R \cdot \mu)$ is positive (which seems as a natural assumption since D can be seen as a percentage dividend of a stock, μ as the expected payoff, and R is a risk-free rate) the whole derivative has negative value.

(c) Recalling that the return is computed as the payoff over price we get:

$$R_u = \frac{D(\mu+\sigma)}{\frac{D(\mu^2-\sigma^2)}{R\mu}} = \frac{R\mu}{(\mu-\sigma)}$$

$$R_d = \frac{D(\mu-\sigma)}{\frac{D(\mu^2-\sigma^2)}{R\mu}} = \frac{R\mu}{(\mu+\sigma)}$$

Hence:

$$E(R) = \frac{1}{2}\frac{R\mu}{(\mu-\sigma)} + \frac{1}{2}\frac{R\mu}{(\mu+\sigma)} = \frac{R\mu}{2}\left(\frac{1}{\mu-\sigma} + \frac{1}{\mu+\sigma}\right)$$

$$= \frac{R\mu}{2}\left(\frac{\mu+\sigma+\mu-\sigma}{\mu^2-\sigma^2}\right) = \frac{R\mu^2}{(\mu^2-\sigma^2)}$$

$$E(R^2) = \frac{1}{2}\frac{R^2\mu^2}{(\mu-\sigma)^2} + \frac{1}{2}\frac{R^2\mu^2}{(\mu+\sigma)^2} = \frac{R^2\mu^2(\mu^2+\sigma^2)}{(\mu^2-\sigma^2)^2}$$

$$Var(R) = E(R^2) - E(R)^2 = \frac{R^2\mu^2(\mu^2+\sigma^2)}{(\mu^2-\sigma^2)^2} - \frac{R^2\mu^4}{(\mu^2-\sigma^2)^2}$$

(d) Recalling the obtained formula for the price

$$S_0 = \frac{D(\mu^2 - \sigma^2)}{R\mu}$$

we can rearrange the variance expression to get:

$$Var(R) = \sigma^2 \frac{D}{S_0^2}$$

Hence:

$$S_0 = \sqrt{\sigma^2 \frac{D}{Var(R)}}$$

and so the price decreases in variance of the return.

4.21

(a) In order to maximize the utility, agent i chooses a strategy $\theta^i = \left(\theta_1^i, \theta_2^i\right)^T$, i.e., agent i buys θ_j^i of asset j subject to his initial endowment. Formally, this can be described by the optimization problems

$$\begin{aligned} &\text{maximize } \frac{1}{3} \ln\left(1 + \theta_1^1\right) + \frac{1}{3} \ln\left(\frac{1}{3} + 1\theta_2^1\right) \\ &\text{subject to } q'\theta^1 \leq 0 \text{ and } w^1 + \theta^1 A > 0 \end{aligned} \tag{4.12}$$

for agent 1 and

$$\begin{aligned} &\text{maximize } \frac{1}{3} \ln\left(1 + \theta_1^2\right) + \frac{1}{3} \ln\left(\frac{1}{3} + \theta_2^2\right) \\ &\text{subject to } q^T\theta^2 \leq 0 \text{ and } w^2 + \theta^2 A > 0 \end{aligned} \tag{4.13}$$

for agent 2. In the optimization problems, the initial endowment the price vector, the probabilities and the payoffs are fixed. The price of the first asset has be normalized to 1. Finally, due to the budget restriction, we replace θ_2^i by $\frac{-\theta_1^i}{q^1}$. This simplifies the optimization problem to a maximization of a function depending on θ_1^i. Differentiating the function with respect to θ_1^i and setting the resulting term equal to 0 gives two equations which can be solved. This gives

$$\theta_1^1 = \frac{q^1 \frac{1}{3} - 1}{2}, \qquad (4.14)$$
$$\theta_1^2 = \frac{q^1 \frac{1}{3} - 1}{2}.$$

In order to “clear away” any excess supply and excess demand, the quantity demanded and the quantity supplied should be equal. In our setup, this means that $\theta_1^1 = -\theta_1^2$ and $\theta_2^1 = -\theta_2^2$ hold. Together with (4.14), these four conditions uniquely determine θ_j^i for $i, j = 1, 2$. Furthermore, combining (4.14) with $\theta_1^1 = -\theta_1^2$, we deduce the explicit form

$$q^1 = \frac{2}{\frac{2}{3}} = 3$$

of the price of asset 2.

(b) In order to exclude bad market behaviour in the sense of arbitrage, it is enough to have existence of an equivalent martingale measure. It is well known, that a tree with three branches defines an incomplete model, i.e., either there is no martingale measure or there is more than one and hence an infinitive number. In our setting, an equivalent martingale measure is a vector $\pi = (\pi_1, \pi_2, \pi_3)^T$ where $\pi_i > 0$ and such that $\pi^T A = q$ holds, i.e.,

$$\begin{bmatrix} 1 \; 3 \end{bmatrix} = [\pi_1 \quad \pi_2 \quad \pi_3] \begin{bmatrix} 1 \; 0 \\ 0 \; 1 \\ 0 \; 1 \end{bmatrix}$$

$$\pi_1 = 1$$
$$\pi_2 + \pi_3 = 3$$
$$\sum_s \pi_s = 4$$

Risk neutral probabilities are then

$$\pi_s^* = \frac{\pi_s}{\sum_s \pi_s} = \frac{\pi_s}{4}$$

Then,

$$\pi_1^* = \frac{1}{4}$$
$$\pi_2^* + \pi_3^* = \frac{\pi_2}{4} + \frac{\pi_3}{4}$$

(c) Aggregate endowment:

$$w = [2 \quad 5 + \frac{1}{3} \quad 5 + \frac{1}{3}]$$

Likelihood ratio is then

$$l_s = \frac{\pi_s^*}{p_s}$$

where $p_s = 1/3$, for all $s = 1, 2, 3$. Then we have

$$l_1 = \frac{1/4}{1/3} = \frac{3}{4}$$

$$l_2 + l_3 = \frac{\frac{\pi_2 + \pi_3}{4}}{1/3} = \frac{3(\pi_2 + \pi_3)}{4}$$

As we see

$$\pi_1 \leq \max\{\pi_2, \pi_3\}.$$

This implies for all possible likelihood ratio processes that would give the correct prices, either l_2 or l_3 must be higher than l_1, and state 2 and state 3 has higher aggregate endowment.

4.22

(a) Since there are only two assets but three states the rank of A is ≤ 2. Thus, the market is not complete.
(b) Yes, we can find

$$q = \frac{1}{3}A_1 + \frac{1}{3}A_2 + \frac{1}{3}A_3$$

(c) The vector of marginal utilities is:

$$\nabla U(w) = \begin{pmatrix} \frac{1}{5} \\ \frac{1}{3} \\ \gamma \end{pmatrix}$$

$$\text{FOC} \quad q = (1, 2) = \left(\tfrac{1}{5} \ \tfrac{1}{3} \ \gamma\right) \begin{pmatrix} 1 & 4 \\ 1 & 2 \\ 1 & 0 \end{pmatrix} = \left(\tfrac{8}{15} + \gamma \ \ \tfrac{22}{15}\right)$$

From here we can solve for γ:

$$\gamma = \frac{3}{15}$$

4.23

(a) The utility function is given by:

$$\begin{aligned} U &= E\left((c-1) - \frac{\gamma}{2}(c-1)^2\right) \\ &= \frac{1}{2}\left((c_u-1) - \frac{\gamma}{2}(c_u-1)^2\right) + \frac{1}{2}\left((c_d-1) - \frac{\gamma}{2}(c_d-1)^2\right) = \frac{1}{2}u(c_u) + \frac{1}{2}u(c_d) \end{aligned}$$

In order for the function to be strongly monotonic, it would have to satisfy:

$$c > c' \; implies \; U(c) > U(c')$$

In the case of this function it holds that:

$$\begin{bmatrix} u(c_u) \\ u(c_d) \end{bmatrix} > \begin{bmatrix} u(c'_u) \\ u(c'_d) \end{bmatrix} \; implies \; U(c) > U(c')$$

but the derivative of w.r.t. c has an ambiguous sign:

$$\frac{\partial u(c)}{\partial c} = 1 - \gamma c + \gamma$$

i.e. it is obtained that the whole function U is strongly monotonic for $c < \frac{1+\gamma}{\gamma}$

(b)

$$\begin{aligned} U &= E\left((c-1) - \frac{\gamma}{2}(c-1)^2\right) \\ &= \frac{1}{2}\left((c_u-1) - \frac{\gamma}{2}(c_u-1)^2\right) + \frac{1}{2}\left((c_d-1) - \frac{\gamma}{2}(c_d-1)^2\right) \\ &= \left(\frac{1}{2}c_u - \frac{1}{2}\right) - \frac{\gamma}{4}(c_u^2 - 2c_u + 1) + \left(\frac{1}{2}c_d - \frac{1}{2}\right) - \frac{\gamma}{4}(c_d^2 - 2c_d + 1) \\ &= \frac{1}{2}c_u + \frac{1}{2}c_d - 1 - \frac{\gamma}{4}(c_u^2 + c_d^2) + \frac{\gamma}{2}c_u + \frac{\gamma}{2}c_d - \frac{\gamma}{2} \\ &= E(c) - 1 - \frac{\gamma}{2}E(c^2) + \gamma E(c) - \frac{\gamma}{2} \\ &= E(c) - 1 - \left(\frac{\gamma}{2}E(c^2) - \frac{\gamma}{2}E(c)^2 + \frac{\gamma}{2}E(c)^2\right) + \gamma E(c) - \frac{\gamma}{2} \\ &= (1+\gamma)E(c) - \left(1 + \frac{\gamma}{2}\right) - \frac{\gamma}{2}E(c)^2 - Varc(c) \end{aligned}$$

(c) The equilibrium is determined as the solution to the following system of equations:

$$\begin{aligned}\theta_0, \theta_1 &= argmax\ U \\ c_u &= \theta_0 q_0 R + \theta_1 S(\mu + \sigma) \\ c_d &= \theta_0 q_0 R + \theta_1 S(\mu - \sigma) \\ \theta_0 q_0 + \theta_1 S &= S\end{aligned}$$

And the market clears:

$$\theta_0 = 0\ and\ \theta_1 = 1$$

Hence:

$$\theta_0 = \frac{S}{q_0}(1 - \theta_1)$$

$$c_u = S(1 - \theta_1)R + \theta_1 S(\mu + \sigma)$$

$$c_d = S(1 - \theta_1)R + \theta_1 S(\mu - \sigma)$$

$$\begin{aligned}U = &\frac{1}{2}\left((S(1 - \theta_1)R + \theta_1 S(\mu + \sigma) - 1) - \frac{\gamma}{2}(S(1 - \theta_1)R + \theta_1 S(\mu + \sigma) - 1)^2\right) \\ &+ \frac{1}{2}\left((S(1 - \theta_1)R + \theta_1 S(\mu - \sigma) - 1) - \frac{\gamma}{2}(S(1 - \theta_1)R + \theta_1 S(\mu - \sigma) - 1)^2\right)\end{aligned}$$

Therefore, from the first order condition of this problem we can get the price of the risky asset, S:

$$\begin{aligned}\frac{\partial U}{\partial \theta_1} = &(-S \cdot R + S(\mu + \sigma)) - \gamma(S(1 - \theta_1)R + \theta_1 S(\mu + \sigma) - 1)(-S \cdot R + S(\mu + \sigma)) \\ &+ (-S \cdot R + S(\mu - \sigma)) - \gamma(S(1 - \theta_1)R + \theta_1 S(\mu - \sigma) - 1)(-S \cdot R + S(\mu - \sigma)) = 0\end{aligned}$$

Substituting the market clearing condition we obtain:

$$\begin{aligned}(-S \cdot R + S(\mu + \sigma)) - \gamma(S(\mu + \sigma) - 1)(-S \cdot R + S(\mu + \sigma) + (-S \cdot R + S(\mu - \sigma)) \\ - \gamma(S(\mu - \sigma) - 1)(-S \cdot R + S(\mu - \sigma)) = 0\end{aligned}$$

After rearranging this expression we obtain the price of the risky asset as:

$$S = \frac{(\mu - R)(1 + \gamma)}{\gamma(\mu^2 + \sigma^2 - \mu R)}$$

4.24 In order to derive the competitive equilibrium in the three cases (a), (b) and (c) we first derive some general equations:

$$\max_{c^i} \sum_{s=0}^{S} \alpha_s^i \ln(c_s^i), \text{ s.t. } \sum_{s=0}^{S} \pi_s c_s^i = \sum_{s=0}^{S} \pi_s w_s^i, \quad i = 1, \dots, I, \tag{4.15}$$

$$\sum_{i=1}^{I} c_s^i = \sum_{i=1}^{I} w_s^i, \quad s = 0, \dots, S, \tag{4.16}$$

with $\pi_0 = 1$. The Lagrangian function of investor i is

$$\mathcal{L}^i(c^i) = \sum_{s=0}^{S} \alpha_s^i \ln(c_s^i) - \lambda^i \sum_{s=0}^{S} \pi_s (c_s^i - w_s^i)$$

and the FOCs are

$$\frac{\partial \mathcal{L}^i}{\partial c_s^i} = \frac{\alpha_s^i}{c_s^i} - \lambda^i \pi_s = 0, \quad s = 0, \dots, S.$$

Solving it with respect to c_s^i, we get:

$$c_s^i = \frac{\alpha_s^i}{\lambda^i \pi_s}, \quad s = 0, \dots, S.$$

Plugging c_s^i into the budget constraint (4.15), we have:

$$\sum_{s=0}^{S} \pi_s w_s^i = \sum_{s=0}^{S} \pi_s c_s^i = \sum_{s=0}^{S} \pi_s \frac{\alpha_s^i}{\lambda^i \pi_s} = \frac{1}{\lambda^i} \sum_{s=0}^{S} \alpha_s^i = \frac{1}{\lambda^i}.$$

Therefore,

$$\lambda^i = \frac{1}{\sum_{s=0}^{S} \pi_s w_s^i},$$

and subsequently

$$c_s^i = \frac{\alpha_s^i}{\pi_s} \sum_{l=0}^{S} \pi_l w_l^i. \tag{4.17}$$

The state prices π may be found from the market clearing conditions (4.16). By Walras Law, it is enough to consider S out of $S + 1$ consumption markets.

(a) The competitive equilibrium is described by the following system of equations:

$$\max_{c^1} U^1(c^1) = \left(\ln(c_0^1) + 2\ln(c_1^1)\right), \text{ s.t. } c_0^1 + \pi_1 c_1^1 = 1 + \pi_1,$$

$$\max_{c^2} U^2(c^2) = \left(2\ln(c_0^2) + \ln(c_1^2)\right), \text{ s.t. } c_0^2 + \pi_1 c_1^2 = 1 + \pi_1,$$

$$c_0^1 + c_0^2 = 2,$$

$$c_1^1 + c_1^2 = 2.$$

Applying formula (4.17), we obtain that

$$c_0^1 = \frac{1}{3\pi_0}(\pi_0 + \pi_1), \quad c_1^1 = \frac{2}{3\pi_1}(\pi_0 + \pi_1),$$

$$c_0^2 = \frac{2}{3\pi_0}(\pi_0 + \pi_1), \quad c_1^2 = \frac{1}{3\pi_1}(\pi_0 + \pi_1).$$

By Walras' Law, market clearing in any of two states is sufficient, e.g.

$$c_0^1 + c_0^2 = \frac{1}{3\pi_0}(\pi_0 + \pi_1) + \frac{2}{3\pi_0}(\pi_0 + \pi_1) = 2,$$

resulting in $\pi_1 = \pi_0 = 1$. Therefore,

$$c_0^1 = \frac{2}{3}, \quad c_1^1 = \frac{4}{3},$$

$$c_0^2 = \frac{4}{3}, \quad c_1^2 = \frac{2}{3}.$$

Comment $\pi_1 = 1$ means that investors are willing to pay price 1 in the first period for 1 unit of the consumption good in the second period.

From agents utility functions, we see that investor 1 prefers consumption in the period 2 over the consumption in the period 1, while investor 2 prefers consumption in the period 1 over the consumption in the period 2. Therefore, in equilibrium we would expect that investor 1 consumes more in the second period than in the first period, while investor 2 consumes more in the first period than in the second period. This is exactly what we find by solving the equations.

(b) The equilibrium is described by the following system of equations:

$$\max_{c^1} U^1(c^1) = \left(\ln(c_1^1) + \ln(c_2^1)\right), \text{ s.t. } \pi_1 c_1^1 + \pi_2 c_2^1 = \pi_1,$$

$$\max_{c^2} U^2(c^2) = \left(\ln(c_1^2) + \ln(c_2^2)\right), \text{ s.t. } \pi_1 c_1^2 + \pi_2 c_2^2 = \pi_2,$$

$$c_1^1 + c_1^2 = 1,$$
$$c_2^1 + c_2^2 = 1.$$

Applying formula (4.17), we obtain that

$$c_0^1 = 0, \quad c_1^1 = \frac{1}{2}, \quad c_2^1 = \frac{\pi_1}{2\pi_2},$$
$$c_0^2 = 0, \quad c_1^2 = \frac{\pi_2}{2\pi_1}, \quad c_2^2 = \frac{1}{2}.$$

Plugging the above expressions into the market clearing condition for the state 1,

$$c_1^1 + c_1^2 = \frac{1}{2} + \frac{\pi_2}{2\pi_1} = 1,$$

we conclude that $\pi_1 = \pi_2$. Therefore,

$$c_1^1 = c_2^1 = c_1^2 = c_2^2 = \frac{1}{2}.$$

Comment $\pi_1 = \pi_2$ means that investors are willing to pay the same price in the first period to get 1 unit of the consumption good in either state 1 or 2 of the second period.

From agents utility functions, we see that investors are indifferent between consumption in states 1 and 2. Moreover, they have homogeneous beliefs regarding the probabilities of state 1 and 2 to occur ($p_1 = p_2 = 0.5$). Therefore, in equilibrium we would expect that investors have the same consumption plans, $c^1 = c^2$ and they consume the same amount independent of the state, $c_1^1 = c_2^1$, $c_1^2 = c_2^2$. This is exactly what we find by solving the equations.

(c) The equilibrium is described by the following system of equations:

$$\max_{c^1} U^1(c^1) = \left(2\ln(c_1^1) + \ln(c_2^1)\right), \text{ s.t. } \pi_1 c_1^1 + \pi_2 c_2^1 = \pi_1 + \pi_2,$$

$$\max_{c^2} U^2(c^2) = \left(\ln(c_1^2) + 2\ln(c_2^2)\right), \text{ s.t. } \pi_1 c_1^2 + \pi_2 c_2^2 = \pi_1 + \pi_2,$$

$$c_1^1 + c_1^2 = 2,$$
$$c_2^1 + c_2^2 = 2.$$

Applying formula (4.17), we obtain that

$$c_0^1 = 0, \quad c_1^1 = \frac{2}{3\pi_1}(\pi_1 + \pi_2), \quad c_2^1 = \frac{1}{3\pi_2}(\pi_1 + \pi_2),$$
$$c_0^2 = 0, \quad c_1^2 = \frac{1}{3\pi_1}(\pi_1 + \pi_2), \quad c_2^2 = \frac{2}{3\pi_2}(\pi_1 + \pi_2).$$

Plugging it into the market clearing for the state 1, we find that $\pi_1 = \pi_2$. Hence,

$$c_1^1 = \frac{4}{3}, \quad c_2^1 = \frac{2}{3},$$
$$c_1^2 = \frac{2}{3}, \quad c_2^2 = \frac{4}{3}.$$

Comment Investors have heterogeneous beliefs regarding the probabilities of state 1 and 2 to occur ($p_1^1 = 2/3$, $p_2^1 = 1/3$, $p_1^2 = 1/3$, $p_2^2 = 2/3$). Therefore, in equilibrium we would expect that investor 1 consumes more in state 1 than in state 2, while investor 2 consumes more in state 2 than in state 1. This is exactly what we find by solving the equations.

(d) The motive for trade in (a) is time preference (saving vs. spending), in (b) risk preference (insurance) and in (c) heterogeneous beliefs (betting).

4.25

We need to prove:

If there exists a $\pi \in \mathbb{R}^S_{++}$ such that $q_k = \sum_{s=1}^S A_s^k \pi_s$, $k = 0, \dots, K$, then there is no $\theta \in \mathbb{R}^{K+1}$ such that $\begin{bmatrix} -q' \\ A \end{bmatrix} \theta > 0$.

We start from $q_k = \sum_{s=1}^S A_s^k \pi_s$, i.e., $\pi' A = q'$. Suppose that there exists an arbitrage opportunity, i.e., there is some $\theta \in \mathbb{R}^{K+1}$ with $\begin{bmatrix} -q' \\ A \end{bmatrix} \theta > 0$. Then, in particular, the first line of this expression must be positive, i.e., $q'\theta < 0$. We put this together with the above formula $q' = \pi' A$ and get $\pi' A\theta < 0$. Since π is positive in all components, this means that $A\theta < 0$, but this conflicts with the assumed positivity of $\begin{bmatrix} -q' \\ A \end{bmatrix} \theta$. Hence, our assumption that there exists an arbitrage opportunity must be wrong.

4.26

(a) As a result of the FTAP we know, that if there is no arbitrage in the sense of strictly monotonic utility, there exists a strictly positive normalized state.

In the case where the number of states is equal to the number of assets this exercise is quite easy to solve: if R has full rank then $R\pi^* = R^f \cdot 1$ implies that

$\pi^* = R^{-1}R^f \cdot 1$. The problem has a unique solution, which we can investigate and see, if it is strictly positive. Thus, if $\pi^* > 0$ then there is no arbitrage via the FTAP.

In our case the problem is more complicated, because we have more states than assets. The problem $R\pi^* = R^f \cdot 1$ has an infinite amount of solutions. The question is if we can find one (or more) that is strictly positive. That means we still need to check, if there is a $\pi^* > 0$ such that $R\pi^* = R^f \cdot 1$. Thus, we numerically solve the set of linear equations subject to the positivity (no-arbitrage) constraint.

To do this we solve the following minimization problem:

$$\min_{\pi^*} \frac{1}{2} \| R\pi^* - R^f \cdot 1 \|_2^2, \quad s.t.\ \pi^* > 0,\ 1'\pi^* = 1$$

This numerically solves the set of linear equations subject to the normalization and the no-arbitrage constraint. If it turns out that the minimum is indeed reasonably close to zero, then we have a solution and thus there exists no arbitrage.

```
%% Exercise 4_23
% Loading the data, calculating returns
clear all
[data,text] = xlsread('Exercise_4_23.xlsx');
riskyreturn = data(2:end,1:11)./data(1:end-1,1:11);
riskyreturn = transpose(riskyreturn);
riskfree = 1+mean(data(:,12))/100;
[assets,states] = size(riskyreturn);
rfvector = riskfree*ones(assets,1);

%% Strict monotonicity
A = -1*eye(states,states);
b = ones(states,1)*-0.000001;
Aeq = ones(1,states);
beq = 1;
[pistar,resid] = lsqlin(riskyreturn,rfvector,A,b,Aeq,beq)
```

The minimum is in fact very close to zero and thus no arbitrage exists in this example.

(b) Since in (a) we have found a solution such that $\forall s, \pi_s^* > 0$, it follows that there exists one with $\forall s, \pi_s^* \geq 0$. Further from the normalization constraint $1'\pi^* = 1$, it follows that $\sum_s \pi_s^* > 0$. This implies that there exists no arbitrage under weak monotonous utility.

(c) We know that we have found a solution in (a) to the problem that is constrained by the normalization and the no-arbitrage constraint. Since the normalization implies that $\sum_s \pi_s^* > 0$, which is the no-arbitrage condition for mean-variance utility, it follows that there must also exist a solution which satisfies this. Therefore, there is no arbitrage possible under mean-variance utility.

4.27

(a) Denote by $\theta = (\theta_B, \theta_S)$ the hedge portfolio:

$$\text{state u:} \qquad \theta_B \cdot R \cdot B + \theta_S \cdot u \cdot S = C_u,$$

$$\text{state d:} \qquad \theta_B \cdot R \cdot B + \theta_S \cdot d \cdot S = C_d.$$

Solving the above equations with respect to θ_B and θ_S, we get

$$\theta_B = \frac{u \cdot C_d - d \cdot C_u}{R \cdot B \cdot (u-d)}, \qquad \theta_S = \frac{C_u - C_d}{(u-d)S}.$$

Therefore, the price of the call option is equal to

$$q_C = \theta_B \cdot B + \theta_S \cdot S = \frac{u \cdot C_d - d \cdot C_u}{R \cdot (u-d)} + \frac{C_u - C_d}{(u-d)}$$
$$= \frac{(u-R) \cdot C_d + (R-d) \cdot C_u}{R \cdot (u-d)}.$$

(b) By the Fundamental Theorem of Asset Prices there exist state prices π, such that $q = A^T \pi$ (assume no-arbitrage market), or equivalently:

$$B = R \cdot B \cdot (\pi_u + \pi_d), \qquad S = S \cdot (u \cdot \pi_u + d \cdot \pi_d).$$

Equations above determine the state prices π in a unique way:

$$\pi_u = \frac{R-d}{R \cdot (u-d)}, \qquad \pi_d = \frac{u-R}{R \cdot (u-d)}.$$

Therefore, the price of the call option may be found as

$$q_C = \frac{1}{R} \mathbb{E}_{\pi} C = \pi_u \cdot C_u + \pi_d \cdot C_d = \frac{(R-d) \cdot C_u + (u-R) \cdot C_d}{R \cdot (u-d)}.$$

(c) The price of the put option is

$$q_P = \frac{(R-d)}{R \cdot (u-d)} \cdot P_u + \frac{(u-R)}{R \cdot (u-d)} \cdot P_d,$$

where P_u is its value in the state u, and P_d is its value in the state d.

$$P_u = \max(K - S \cdot u, 0) = \max(100 - 200, 0) = 0,$$
$$P_d = \max(K - S \cdot d, 0) = \max(100 - 50, 0) = 50,$$

therefore

$$q_P = \frac{0.6 \cdot 0 + 0.9 \cdot 50}{1.1 \cdot 1.5} = 27.27.$$

4.28

(a) The barrier option cannot be hedged since the rank of the matrix $[A, A_3]$ is three. This can be checked by calculating the determinant of the first three rows of the matrix $[A, A_3]$. If this determinant is nonzero, the matrix of these three rows has full rank. Therefore, there is no linear combination of the stock and the bond which can hedge the barrier option.

(b) The price bounds of the barrier option are determined via

$$\overline{q}(y) = \max_{\pi} \pi' y \qquad \text{s.t. } A'\pi = q \text{ and } \pi \geq 0$$
$$\underline{q}(y) = \min_{\pi} \pi' y \qquad \text{s.t. } A'\pi = q \text{ and } \pi \geq 0$$

This problem has to be solved numerically. A solution with Excel can be found online. The solutions are

$$\overline{q}(y) = 1.03 \qquad \underline{q}(y) = 0.76$$

(c) Also this problem has to be solved numerically. A solution with Excel can be found online. A possible solution is

$$\alpha = 0.7 \qquad \alpha^+ = 0.4 \qquad \alpha^- = 1 \qquad \beta = 1.5 \qquad RP = 1.2$$

4.29

(a) Remember the definition of the no-arbitrage price bounds in terms of state prices:

1. agents have **strictly monotonic utilities**:

$$\overline{q}(y) = \max_{\pi >> 0} \pi^T y, \text{ such that } \pi^T A = q^T,$$
$$\underline{q}(y) = \min_{\pi >> 0} \pi^T y, \text{ such that } \pi^T A = q^T,$$

For our example we get:

$$\pi^T A = q^T \iff \begin{cases} \pi_1 + \pi_2 + \pi_3 = 0.9, \\ \pi_1 = 0.25, \end{cases} \iff \begin{cases} \pi_1 = 0.25, \\ \pi_2 + \pi_3 = 0.65. \end{cases}$$

Therefore,

$$\overline{q}(A_3) = \max_{\pi >> 0} (0.25 + 2\pi_2), \text{ such that } \pi_2 + \pi_3 = 0.65,$$

$$\underline{q}(A_3) = \min_{\pi >> 0} (0.25 + 2\pi_2), \text{ such that } \pi_2 + \pi_3 = 0.65.$$

Clearly $\overline{q}(A_3) = 1.55$ and is achieved when

$$\pi_2 = 0.65 - \frac{1}{n}, \quad \pi_3 = \frac{1}{n}, \quad n \to \infty,$$

while $\underline{q}(A_3) = 0.25$ and is achieved when

$$\pi_2 = \frac{1}{n}, \quad \pi_3 = 0.65 - \frac{1}{n}, \quad n \to \infty.$$

(b) The bounds for the security price may be also found via dominating and dominated portfolios:

$$\overline{q}(A_4) = \min_{\theta} q^T \theta, \text{ such that } A\theta \geq A_4,$$

$$\underline{q}(A_4) = \min_{\theta} q^T \theta, \text{ such that } A\theta \leq A_4.$$

If θ is a dominating portfolio, then

state 1:	$\theta_1 + \theta_2 \geq 1,$
state 2:	$\theta_1 \geq 0,$
state 3:	$\theta_1 \geq 1,$

and its value is $0.9\theta_1 + 0.25\theta_2$. Clearly, the minimum is achieved when $\theta_1 = 1$ and $\theta_2 = 0$, therefore, $\overline{q}(A_4) = 0.9$.

Analogously, if θ is a dominated portfolio, then

state 1:	$\theta_1 + \theta_2 \leq 1,$
state 2:	$\theta_1 \leq 0,$
state 3:	$\theta_1 \leq 1,$

and its value is $0.9\theta_1 + 0.25\theta_2 = 0.25(\theta_1 + \theta_2) + 0.65\theta_1 \le 0.25 + 0.65\theta_1$. Therefore, the maximum portfolio value is achieved when $\theta_1 = 0$ and $\theta_2 = 1$, so $\overline{q}(A_4) = 0.25$.

4.30

First, we prove that if there is no $\theta \ge 0$ such that $q\theta \le 0$ and $A\theta > 0$, then $q \gg 0$. That is, under positive payoffs and shortsale constraints, no-arbitrage implies positive prices.

Suppose there exists $q^k \le 0$ (i.e., the price for asset k is less than or equal to 0), then we can find $\theta = [0, 0, \ldots, 1, 0, \ldots, 0]$, where only the k-th element is 1, and all other elements are 0. In this case, there exists $\theta \ge 0$, so that $q'\theta \le 0$ $A\theta > 0$, which conflicts with the no-arbitrage condition. Hence, with frictions, no-arbitrage implies positive prices.

Second, we prove that if $q \gg 0$, then there is no $\theta \ge 0$ such that $q'\theta \le 0$ and $A\theta > 0$. That is, positive prices imply no-arbitrage under positive payoffs and shortsale constraints.

Assume there is an arbitrage opportunity, i.e., we can find $\theta \ge 0$ such that $q'\theta \le 0$ and $A\theta > 0$. Then θ must be less than or equal to 0, in order to satisfy $q'\theta \le 0$. This conflicts with $q \gg 0$ (i.e., positive price). So under shortsale constraints, positive prices are arbitrage-free.

4.31

(a) Since A is a square matrix, the financial market is complete if and only if the determinant of A is not equal to zero. We have:

$$\det(A) = -2 \neq 0,$$

hence, the financial market is complete.

(b) Both investors have Cobb-Douglas utilities, hence we can apply the result (4.17) of the exercise (3.4):

$$c_1^i = \frac{p\left(\pi_1 w_1^i + \pi_2 w_2^i\right)}{\pi_1}, \quad c_2^i = \frac{(1-p)\left(\pi_1 w_1^i + \pi_2 w_2^i\right)}{\pi_2},$$

where w_s^i is income of agent i in state s from holding asset endowment θ_A^i:

$$w^1 = (0.5, 1)', \quad w^2 = (2, 0)'.$$

From the market clearing in, e.g., state 1 we have:

$$c_1^1 + c_1^2 = 2.5 = \frac{p}{\pi_1}(0.5\pi_1 + \pi_2) + \frac{p}{\pi_1} 2\pi_1,$$

or equivalently,

$$\frac{\pi_2}{\pi_1} = 2.5\left(\frac{1}{p} - 1\right) = 1.$$

Recall, that by Walras Law it is enough to consider market clearing in one state only (market clearing in the other state 2 is satisfied automatically).

Finally, plugging in the values for p and w^i, $i = 1, 2$, into (4.18) we obtain the equilibrium consumption:

$$c^1 = \left(c_1^1, c_2^1\right) = \left(\frac{15}{14}, \frac{3}{7}\right), \quad c^2 = \left(c_1^2, c_2^2\right) = \left(\frac{10}{7}, \frac{4}{7}\right).$$

(c) For the equilibrium prices $q = (q^1, q^2)'$ relation

$$q = A'\pi$$

holds. Using the result $\pi_2 = \pi_1$ for the state prices from part (a), we therefore find that

$$\frac{q^2}{q^1} = \frac{4}{3}.$$

The equilibrium asset allocation can be found from the equilibrium consumption:

$$c^i = A\theta^i,$$

or equivalently,

$$\theta^i = A^{-1}c^i.$$

Plugging in the values for consumption obtained in (a), we get:

$$\theta^1 = \frac{3}{7}(1, 1)', \quad \theta^2 = \frac{4}{7}(1, 1)'.$$

(d) Recall that asset allocation λ^i is given by:

$$\lambda_k^i = \frac{q^k \theta_k^i}{\sum_{l=1,2} q^l \theta_l^i},$$

and the asset returns are:

$$R = A \cdot diag(1/q) = \begin{pmatrix} A_1^1/g^1 & A_1^2/q^1 \\ A_2^1/q^1 & A_2^2/q^2 \end{pmatrix}.$$

For agent 1 we have:

$$\lambda_1^1 = \frac{q^1 \frac{3}{7}}{q^1 \frac{3}{7} + q^2 \frac{3}{7}} = \frac{q^1 \frac{3}{7}}{q^1 \frac{3}{7} + q^1 \frac{4}{3} \cdot \frac{3}{7}} = \frac{3}{7},$$

$$\lambda_2^1 = \frac{q^2 \frac{3}{7}}{q^1 \frac{3}{7} + q^2 \frac{3}{7}} = \frac{q^1 \frac{4}{3} \cdot \frac{3}{7}}{q^1 \frac{3}{7} + q^1 \frac{4}{3} \cdot \frac{3}{7}} = \frac{4}{7},$$

where we used that $q^2 = \frac{4}{3} q^1$. Similarly, for agent 2 we obtain:

$$\lambda^2 = \frac{1}{7}(3, 4)'.$$

Finally, for the asset returns we get:

$$R = \frac{1}{q^1} \begin{pmatrix} 0.5 & 1.5 \\ 1 & 0 \end{pmatrix}.$$

(e) The 2×2 matrix of factor loadings β is implicitly defined by $(R_s^k) = (F_s^f)(\beta_k^f)'$. Therefore, we obtain:

$$(\beta_k^f) = (F^{-1} R)' = \frac{1}{6} \begin{pmatrix} 3 & 5 \\ -3 & 3 \end{pmatrix}.$$

(f) For the factor prices q^F we have:

$$q^F = F' \pi = \frac{1}{2} \begin{pmatrix} -1 \\ 3 \end{pmatrix}.$$

Denote by θ_F^i the factors' portfolio of agent i. The payoff of the factors' portfolio should be the same as the payoffs of equilibrium asset allocation θ^i:

$$A\theta^i = F\theta_F^i.$$

Therefore, we obtain:

$$\theta_F^i = F^{-1} A \theta^i = \frac{1}{2} \begin{pmatrix} -1 & 2 \\ 1 & 2 \end{pmatrix} \cdot \frac{1}{2} \begin{pmatrix} 1 & 4 \\ 2 & 0 \end{pmatrix},$$

or after plugging the values for θ^i:

$$\theta_F^1 = \frac{3}{28}(-1, 9)', \quad \theta_F^2 = \frac{1}{7}(-1, 9)'.$$

4.32 The equilibrium of this economy is defined as the solution to the following equation system:

$$c_0^i, c_1^i = argmax\ U^i$$

s.t.

$$\begin{aligned} c_0^i &= w_0^i - \theta_0^i q_0 \\ c_1^i &= w_0^i(1+g) + \theta_0^i q_0 R_f \end{aligned}$$

And market clears:

$$\sum_{i=1}^{I} \theta_0^i = 0$$

Hence:

$$\begin{aligned} U^i &= \ln(w_0^i - \theta_0^i q_0) + \frac{1}{1+\delta} \ln(w_0^i(1+g) - \theta_0^i q_0 R_f) \\ \frac{\partial U^i}{\partial \theta_0^i} &= \frac{-q}{(w_0^i - \theta_0^i q_0)} + \frac{1}{1+\delta} \frac{q_0 R_f}{(w_0^i(1+g) + \theta_0^i q_0 R_f)} = 0 \\ \frac{1}{(w_0^i - \theta_0^i q_0)} &= \frac{R_f}{(1+\delta)(w_0^i(1+g) + \theta_0^i q_0 R_f)} \\ \theta_0^i(q_0 R_f(1+\delta) + q_0 R_f) &= w_0^i(R_f - (1+g)(1+\delta)) \\ \theta_0^i &= w_0^i \frac{(R_f - (1+g)(1+\delta))}{q_0 R_f (2+\delta)} \end{aligned}$$

Summing this expression over all agents and substituting the market clearing condition we get:

$$\begin{aligned} \sum_{i=1}^{I} \theta_0^i = \left(\sum_{i=1}^{I} w_0^i \right) \frac{(R_f - (1+g)(1+\delta))}{q_0 R_f (2+\delta)} &= 0 \\ \frac{(R_f - (1+g)(1+\delta))}{q_0 R_f (2+\delta)} &= 0 \end{aligned}$$

and assuming that q_0, R_f and δ are all non-negative it follows that:

$$R_f - (1+g)(1+\delta) = 0$$
$$R_f = (1+g)(1+\delta)$$

4.33

(a) By the Fundamental Theorem of Asset Pricing (FTAP), the absence of arbitrage is equivalent to the existence of state prices (risk-neutral probabilities) π^*, such that

$$R^f = \pi_u^* u + \pi_d^* d,$$
$$\pi_u^* + \pi_d^* = 1.$$

Hence, we obtain:

$$\pi_u^* = \frac{R^f - d}{u - d}, \quad \pi_d^* = \frac{u - R^f}{u - d}.$$

(b) The utility maximization problem of the representative agent for the complete market is:

$$\max_{c_u, c_d} U(c_u, c_d) = p \log(c_u) + (1-p)\log(c_d),$$
$$\text{s.t. } \pi_u c_u + \pi_d c_d = \pi_u u + \pi_d d,$$

where π_u and π_d are the state prices. The Lagrange function is:

$$\mathcal{L}(c_u, c_d) = p\log(c_u) + (1-p)\log(c_d) - \lambda\left(\pi_u c_u + \pi_d c_d - (\pi_u u + \pi_d d)\right),$$

and the FOCs are:

$$\frac{\partial \mathcal{L}}{\partial c_u} = \frac{p}{c_u} - \lambda \pi_u = 0,$$
$$\frac{\partial \mathcal{L}}{\partial c_d} = \frac{1-p}{c_d} - \lambda \pi_d = 0.$$

Excluding λ, we obtain:

$$\frac{p}{c_u \pi_u} = \lambda = \frac{1-p}{c_d \pi_d},$$

or equivalently,

$$\frac{\pi_u}{\pi_d} = \frac{p}{1-p}\frac{c_d}{c_u}.$$

From the market clearing conditions in the second period we conclude that $c_u = u$ and $c_d = d$. Therefore, for the risk-neutral probabilities (normalized state prices) we have:

$$\pi_u^* = \frac{pd}{(1-p)u + pd}, \qquad \pi_d^* = \frac{(1-p)u}{(1-p)u + pd}.$$

(c) Combining the results from parts (a) and (b), we have:

$$\frac{R^f - d}{u - d} = \pi_u^* = \frac{pd}{(1-p)u + pd}.$$

Consequently,

$$R^f = \frac{ud}{(1-p)u + pd}.$$

(d) This problem can be solved analogously to (a)–(c) for an utility function $U(c) = \frac{u^{1-\gamma}}{1-\gamma}$. The results are as follows:

$$\pi_u^* = \frac{pu^{-\gamma}}{(1-p)d^{-\gamma} + pu^{-\gamma}}, \qquad \pi_d^* = \frac{(1-p)d^{-\gamma}}{(1-p)d^{-\gamma} + pu^{-\gamma}},$$

$$R^f = \frac{pu^{1-\gamma} + (1-p)d^{1-\gamma}}{(1-p)d^{-\gamma} + pu^{-\gamma}}.$$

Note that the logarithmic utility considered in (a)–(c) is a special case of the CRRA utility with $\gamma = 1$.

4.34 The optimization problem of the representative agent is:

$$\max_c u(c_0) + \delta \sum_{s=1}^{S} p_s u(c_s) \quad \text{s.t. } w_0 + \sum_s \pi_s w_s = c_0 + \sum_s \pi_s c_s.$$

By substituting the budget constraint

$$c_0 = w_0 + \sum_s \pi_s w_s - \sum_s \pi_s c_s$$

into the investor's utility function, we can rewrite the optimization problem of the representative investor:

$$\max_{c} u\left(w_0 + \sum_s \pi_s w_s - \sum_s \pi_s c_s\right) + \delta \sum_{s=1}^{S} p_s u(c_s).$$

The FOCs are:

$$u'(c_0)(-\pi_s) + \delta p_s u'(c_s) = 0, \quad s = 1, \ldots, S.$$

For the likelihood ratio process ℓ we obtain that it is equal to the marginal rate of substitution:

$$\ell_s = \frac{\pi_s^*}{p_s} = \frac{\pi_s}{p_s \sum_{l=1}^{S} \pi_l} = \frac{1}{p_s} \frac{\frac{\delta p_s u'(c_s)}{u'(c_0)}}{\sum_{l=1}^{S} \frac{\delta p_l u'(c_l)}{u'(c_0)}} = \frac{u'(c_s)}{\mathbb{E}_P u'(c_l)}. \tag{4.18}$$

Note that

$$\frac{1}{R_f} = \sum_{s=1}^{S} \pi_s,$$

so likelihood ratio (4.18) may be also written as

$$\ell_s = R_f \frac{\pi_s}{p_s} = R_f \frac{\delta u'(c_s)}{u'(c_0)}, \quad s = 1, \ldots, S.$$

(a) Quadratic utility: $u'(c) = 1 - \gamma c$, therefore,

$$\ell_s = \delta R_f \frac{1 - \gamma c_s}{1 - \gamma c_0}, \quad s = 1, \ldots, S.$$

(b) CRRA utility: $u'(c) = c^{-\rho}$, therefore,

$$\ell_s = \delta R_f (c_s / c_0)^{-\rho}, \quad s = 1, \ldots, S.$$

(c) CARA utility: $u'(c) = \alpha e^{-\alpha c}$, therefore,

$$\ell_s = \delta R_f e^{\alpha(c_0 - c_s)}, \quad s = 1, \ldots, S.$$

(d)

$$u'(c) = \begin{cases} \alpha^+ (c - RP)^{\alpha^+ - 1} & \text{if } c > RP, \\ \alpha^- \beta (RP - c)^{\alpha^- - 1} & \text{if } c \le RP, \end{cases}$$

therefore,

$$\ell_s = \delta R_f \frac{\alpha^+(c_s - RP)^{\alpha^+-1} I(c_s > RP) + \alpha^- \beta (RP - c_s)^{\alpha^- -1} I(c_s \leq RP)}{\alpha^+(c_0 - RP)^{\alpha^+-1} I(c_0 > RP) + \alpha^- \beta (RP - c_0)^{\alpha^- -1} I(c_0 \leq RP)}.$$

(e)

$$u'(c) = \begin{cases} 1 - 2\alpha^+(c - RP) & \text{if } c > RP, \\ \beta\left[1 - 2\alpha^-(RP - c)\right] & \text{if } c \leq RP, \end{cases}$$

therefore,

$$\ell_s = \delta R_f \frac{[1 - 2\alpha^+(c_s - RP)] I(c_s > RP) + \beta\left[1 - 2\alpha^-(RP - c_s)\right] I(c_s \leq RP)}{[1 - 2\alpha^+(c_0 - RP)] I(c_0 > RP) + \beta\left[1 - 2\alpha^-(RP - c_0)\right] I(c_0 \leq RP)}.$$

4.35

From the derivation of the CAPM we know

$$l_s = \frac{1 - \gamma c_s}{1 - \gamma \mathbb{E}(c)} = \frac{1}{1 - \gamma \mathbb{E}(R^M)} - \frac{\gamma}{1 - \gamma \mathbb{E}(R^M)} R_s^M$$

and

$$\frac{\gamma}{1 - \gamma \mathbb{E}(R^M)} = \frac{\mathbb{E}\left(R^M - R^f\right)}{\text{VAR}(R^M)}$$

Plug this into l_s, take the expectation of both sides of the equation and use $\mathbb{E}(l) = 1$, we obtain

$$\begin{aligned}
1 = \mathbb{E}(l) &= \frac{1}{1 - \gamma \mathbb{E}(R^M)} - \frac{\mathbb{E}\left(R^M - R^f\right)}{\text{VAR}(R^M)} \mathbb{E}(R^M) \\
\frac{1}{1 - \gamma \mathbb{E}(R^M)} &= 1 + \frac{\mathbb{E}\left(R^M - R^f\right)}{\text{VAR}(R^M)} \mathbb{E}(R^M) \\
l_s &= 1 + \frac{\mathbb{E}\left(R^M - R^f\right)}{\text{VAR}(R^M)} \mathbb{E}(R^M) - \frac{\mathbb{E}\left(R^M - R^f\right)}{\text{VAR}(R^M)} R_s^M \\
&= a - b R_s^M
\end{aligned}$$

with

$$a = 1 + \frac{\mathbb{E}\left(R^M - R^f\right)}{\text{VAR}(R^M)} \mathbb{E}(R^M)$$

$$b = \frac{\mathbb{E}\left(R^M - R^f\right)}{\text{VAR}(R^M)}$$

Since all states are equally likely we have $p_s = p = 1/S$. Therefore, we obtain:

$$\pi_s^* = l_s p_s = \frac{a - bR_s^M}{S} =: \pi_s^{CAPM}$$

We should find now a $\overline{\pi}_s^*$, which minimizes the distance to the CAPM and for which no arbitrage holds, i.e.

$$\min_{\overline{\pi}_s^*} = \sum_s (\overline{\pi}_s^* - \pi_s^{CAPM})^2$$

$$\text{s.t. } \sum_s \overline{\pi}_s^* = 1$$

$$\sum_s \overline{\pi}_s^* (R_s^k - R_f) = 0 \qquad \text{for all } k$$

$$\overline{\pi}_s^* > 0 \qquad \text{for all } s$$

```
%% Exercise 4_32
% Loading the data, calculating returns
clear all
[data,text] = xlsread('Exercise_4_23.xlsx');
riskyreturn = data(2:end,1:11)./data(1:end-1,1:11);
riskyreturn = transpose(riskyreturn);
riskfree = 1+mean(data(:,12))/100;
[assets,states] = size(riskyreturn);
rfvector = riskfree*ones(assets,1);

Erm = mean(riskyreturn(1,:)'); % Market expected return
ExErm = Erm - riskfree; % Market expected excess return
VarRm = var(riskyreturn(1,:));
a = 1+ExErm/VarRm*Erm;
b = ExErm/VarRm;
picapm = (a-b*riskyreturn(1,:)')./states; % CAPM implied norm.
state prices

A = -1*eye(states,states);
b = ones(states,1)*-0.000001;
Aeq = [ones(1,states); riskyreturn];
beq = [1;rfvector];
[pistar,resid] = lsqlin(eye(states),picapm,A,b,Aeq,beq)

%% Relative deviations
deviation = (pistar-picapm)./picapm;
plot(picapm,deviation,'o');
```

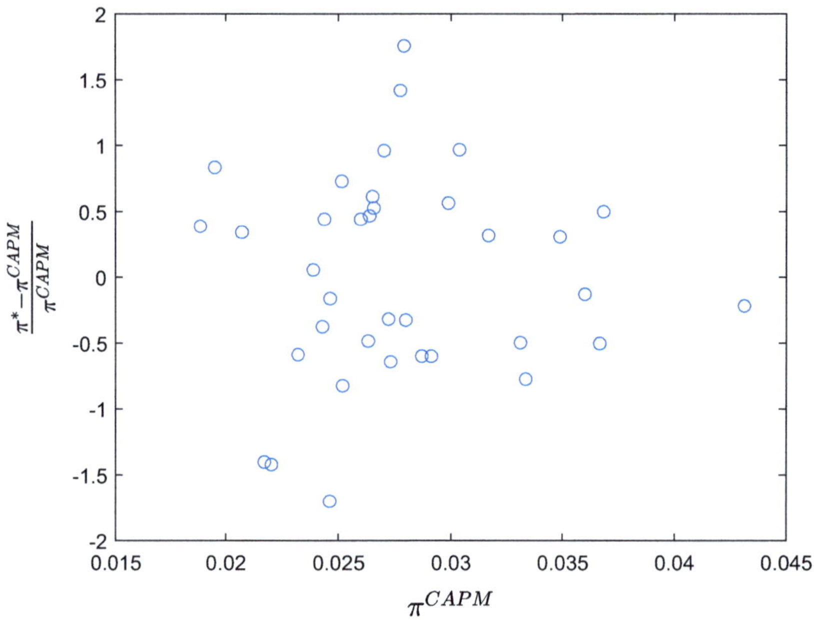

Fig. 4.3 Plot of the relative deviation of the calculated normalized state prices from the ones implied by CAPM

We find that the residual in the minimum point found numerically is quite big. To further investigate this, Fig. 4.3 shows the relative deviation of each π_s^* versus the normalized state prices implied by CAPM (π_s^{CAPM}).

4.36

(a) Look at the Arbitrage Pricing Model:

$$R_t^{e,k} = \beta_k'(f_t - \mathbb{E}(f)) + \epsilon_t^k$$

where ϵ_t^k is white noise. It can be easily checked that $\beta^k = \mathrm{VAR}(f)^{-1}\,\mathrm{cov}(f, R^{e,k})$ is just the OLS estimator of β_k of the APT.

(b) Using $R^f = \mathbb{E}(R^k) + \mathrm{cov}(l, R^k)$, $l = 1 + b'(f - \mathbb{E}(f))$ and $R^{e,k} = R^k - R_f$ it can be written:

$$\begin{aligned}\mathbb{E}(R^{e,k}) &= -\,\mathrm{cov}(1 + b'(f - \mathbb{E}(f)), R^{e,k}) = -b'\,\mathrm{cov}(f, R^{e,k})\\ &= -b\,\mathrm{VAR}(f)\,\mathrm{VAR}(f)^{-1}\,\mathrm{cov}(f, R^{e,k})\\ &= \lambda'\beta\end{aligned}$$

(c) The estimation can be done as follows:
   1. β_k can be estimated be OLS from the following equation:

$$R_t^{e,k} = \beta_k'(f_t - \mathbb{E}(f)) + \epsilon_t^k$$

   where the average is taken for $\mathbb{E}(f)$.
   2. λ can be estimated by OLS out of

$$\mathbb{E}\left(R_t^{e,k}\right) = \lambda'\beta_k + \eta^k$$

   with η^k white noise. For $\mathbb{E}\left(R_t^{e,k}\right)$ the average and for β_k the estimator from the last step is taken.
   3. Since $\lambda = -\operatorname{VAR}(f)b$, b is obtained via

$$b = -\operatorname{VAR}(f)^{-1}\lambda$$

   where the empirical covariance matrix is used for $\operatorname{VAR}(f)$ and the estimated λ from the last step is used.

4.37

(a) Calculate the return of the actual portfolio of the pension fund:

$$\mu_P - R_f = \beta_P(\mu_M - R_f)$$
$$\mu_P = 5.5\%$$

Calculate the sensitivity to the unexpected inflation:

$$\mu_P - R_F = -b_{Inf}^P \cdot 1\% + b_{GDP}^P \cdot 2\%$$

since $\mu_P - R_f = 2.5\%$ and $b_{GDP}^P = 0.2$ we obtain $b_{Inf}^P = -2.1$. In other words if the inflation is high the returns are low.

(b)

$$\mu_P - R_F = -b_{Inf}^P \cdot 1\% + b_{GDP}^P \cdot 2\%$$

with $b_{Inf}^P = 0$ and $b_{GDP}^P = 0$

$$\mu_P = R_f = 3\%$$

(c) The board wants a new portfolio P':

$$\mu_{P'} - R_F = -b_{Inf}^{P'} \cdot 1\% + b_{GDP}^{P'} \cdot 2\%$$

with $\mu_{P'} = 6\%$ and $b^P_{Inf} = 0$ therefore

$$b^{P'}_{GDP} = 1.5$$

4.38

(a) The given economy is the usual exchange economy with logarithmic preferences and complete market. Therefore, we can apply the result of the Exercise 4.24,

$$c^i_s = \frac{\alpha^i_s}{\pi_s} \sum_{l=1,2} \pi_l w^i_l,$$

with $\alpha^1 = 0.75$, $\alpha^2 = 0.25$, $w^1_1 = w^1_2 = 1$ (income from holding asset 1), $w^2_1 = 2$ and $w^2_2 = 0.5$ (income from holding asset 2). By the market clearing in state 1 we have

$$c^1_1 + c^2_1 = w^1_1 + w^2_1,$$

hence, after Plugging the values for α^i and w^i, we obtain:

$$\frac{\pi_2}{\pi_1} = \frac{\sum_{i=1,2}(1-\alpha^i_1)w^i_1}{\sum_{i=1,2}\alpha^i_1 w^i_2} = \frac{(1-0.75)\cdot 1 + (1-0.25)\cdot 2}{0.75\cdot 1 + 0.25\cdot 0.5} = 2.$$

The equilibrium asset prices are then:

$$q = A'\pi = (\pi_1 + \pi_2, 2\pi_1 + 0.5\pi_2)' = \pi_1(3,3)'$$

with $A = \begin{pmatrix} 1 & 2 \\ 1 & 0.5 \end{pmatrix}$ being the matrix of asset payoffs, and the optimal consumption is:

$$c^1 = (\pi_1 \cdot 1 + \pi_2 \cdot 1)\big(0.75/\pi_1, 0.25/\pi_2\big)' = \frac{3}{8}(6,1)',$$

$$c^2 = (\pi_1 \cdot 2 + \pi_2 \cdot 0.25)\big(0.25/\pi_1, 0.75/\pi_2\big)' = \frac{3}{8}(2,3)'.$$

Finally, we find the equilibrium asset allocation from the second period budget constraints:

$$c^i = A\theta^i, \quad i = 1,2,$$

or equivalently,

$$\theta^i = A^{-1}c^i, \quad i = 1,2.$$

Plugging the values for the optimal consumption, we obtain:

$$\theta^1 = \frac{1}{4}(-1, 5)' \text{ and } \theta^2 = \frac{1}{4}(5, -1)'.$$

(b) Let

$$U^R(c_1, c_2) = \gamma \ln(c_1) + (1-\gamma)\ln(c_2)$$

be the utility function of the representative consumer. We need to find the value of the parameter γ, such that the representative agent is able to replicate the prices from (a).

Applying the result (4.19) of the previous section we obtain that

$$\frac{\pi_2}{\pi_1} = \frac{(1-\gamma)w_{A,1}}{\gamma w_{A,2}} = \frac{\gamma}{(1-\gamma)} \frac{1 \cdot 1 + 1 \cdot 2}{1 \cdot 1 + 1 \cdot 0.5} = \frac{2(1-\gamma)}{\gamma},$$

where $w_{A,s}$ is the aggregated endowment of investors in state $s = 1, 2$. From (a) we know that $\frac{\pi_2}{\pi_1} = 2$. Hence, a representative agent with γ, such that

$$\frac{2(1-\gamma)}{\gamma} = 2,$$

or equivalently, $\gamma = 0.5$, generates the same asset prices as the two investors in (a).

(c) The new aggregated endowment is $w_A = (4, 1.5)'$. Using (4.19) we obtain that

$$\frac{\pi_2}{\pi_1} = \frac{(1-\gamma)w_{A,1}}{\gamma w_{A,2}} = \frac{0.5 \cdot 4}{0.5 \cdot 1.5} = \frac{8}{3}.$$

The asset prices are then

$$q^1 = 1 \cdot \pi_1 + 1 \cdot \pi_2 = \frac{11}{3}\pi_1,$$

$$q^2 = 3 \cdot \pi_1 + 0.5 \cdot \pi_2 = \frac{13}{3}\pi_1,$$

with $\frac{q^1}{q^2} = \frac{11}{13}$. Letting $\pi_2 = 2$ as in (b) leads to $\pi_1 = \frac{3}{4}$ and $q = \frac{1}{4}(11, 13)'$.

(d) The new endowments of investors are $w^1 = (1, 1)'$ and $w^2 = (3, 0.5)'$. Therefore, from (4.19) we find that

$$\frac{\pi_2}{\pi_1} = \frac{\sum_{i=1,2}(1-\alpha_1^i)w_1^i}{\sum_{i=1,2}\alpha_1^i w_2^i} = \frac{0.25 * 1 + 0.75 * 3}{0.75 * 1 + 0.25 * 0.5} = \frac{20}{7}.$$

The new asset prices are then:

$$q_1 = 1 \cdot \pi_1 + 1 \cdot \pi_2 = \frac{27}{20}\pi_2,$$

$$q_2 = 3 \cdot \pi_1 + 0.5 \cdot \pi_2 = \frac{31}{20}\pi_2,$$

with $\frac{q_1}{q_2} = \frac{27}{31}$. Letting $\pi_2 = 2$ as in (a) leads to $\pi_1 = 0.7$ and $q = (2.7, 3.1)'$.

4.39

(a) A financial markets equilibrium is a list of portfolio strategies $\theta^{i,*}$, $i = 1, \ldots, I$, and a price system q^k, $k = 1, \ldots, K$, such that for all $i = 1, \ldots, I$:

$$\theta^{i,*} = \arg\max_{\theta^i} \; U^i(c^i), \; \text{s.t.} \begin{cases} q'\theta^i \leq 0, \\ c^i = w^i + A\theta^i, \end{cases} \tag{4.19}$$

and markets clear:

$$\sum_{i=1}^{I} \theta^i = 0. \tag{4.20}$$

(b) It has to be checked that the investors have optimized their portfolios and the markets are cleared.

The market clearing condition can be easily verified by plugging the given values into (4.20). As for the decision problems of agents, (4.19), first we simplify them. Inserting the second period budget constraints into investors' utility functions and plugging the values $q^1 = 1$ and $q^2 = 4$ for the asset prices, we obtain:

$$\max_{\theta^1} \; \ln(\theta_1^1) + \ln(\theta_2^1 + 1), \; \text{s.t} \begin{cases} \theta_1^1 + 4\theta_2^1 \leq 0, \\ c_1^1 = \theta_1^1 \geq 0, \\ c_2^1 = \theta_2^1 + 1 \geq 0, \\ c_3^1 = \theta_2^1 + 2 \geq 0, \end{cases}$$

$$\max_{\theta^2} \; \ln(\theta_2^2 + 1) + \ln(\theta_2^2), \; \text{s.t} \begin{cases} \theta_1^2 + 4\theta_2^2 \leq 0, \\ c_1^2 = \theta_1^2 + 2 \geq 0, \\ c_2^2 = \theta_2^2 + 1 \geq 0, \\ c_3^2 = \theta_2^2 \geq 0, \end{cases}$$

where with $c_s^i \geq 0$ we accounted for non-negative consumption. The utility function of investor 1 is strictly monotonous with respect to his asset holdings θ^1, therefore the first period budget constraint is binding:

$$\theta_1^1 = -4\theta_2^1.$$

The utility maximization problem of investor 1 then simplifies as follows:

$$\max_{\theta_2^1} \ln(-4\theta_2^1) + \ln(1 + \theta_2^1), \text{ s.t. } -1 \leq \theta_2^1 \leq 0. \tag{4.21}$$

From the first order condition of the unconstrained problem,

$$\frac{1}{\theta_2^1} + \frac{1}{1 + \theta_2^1} = 0,$$

we find that $\theta_2^1 = -\frac{1}{2}$. From the first period budget constraint we have $\theta_1^1 = -4\theta_2^1 = 2$. Since the constraint in (4.21) is satisfied, the portfolio $\theta^1 = (2, -0.5)$ is the solution of the decision problem of investor 1.

Similarly we proceed with the optimization problem investor 2. From his utility function we see that investing in asset 1 brings no utility to agent 2. Therefore he invests as few as possible into asset 1. The consumption $c_1^2 = \theta_1^2 + 2$ in state 1 must be nonnegative, hence $\theta_1^2 = -2$. On the other hand, investor 2 would like to invest as much as possible into asset 2, therefore his first period budget constraint is binding:

$$\theta_1^2 + 4\theta_2^2 = 0,$$

or equivalently, $\theta_2^2 = -\theta_1^2/4 = 0.5$. For the given value of θ_2^2 the consumption $c_2^2 = \theta_2^2 + 1$ and $c_3^2 = \theta_2^2$ is also nonnegative. Hence, the portfolio $\theta^2 = (-2, 0.5)'$ is the solution of the utility maximization problem of investor 2. Thereby, $\theta^{1,*}$, $\theta^{2,*}$ and q^* is indeed an equilibrium.

(c) The asset 3 can be replicated by the other two assets if and only if the column rank of the matrix $[A, A^3]$ is equal to the column rank of the matrix A. Since the matrix

$$[A, A^3] = \begin{pmatrix} 1 & 0 & 0 \\ 0 & 1 & 0 \\ 0 & 1 & 1 \end{pmatrix}$$

has full rank ($det([A, A^3]) \neq 0$), we conclude that asset 3 cannot be duplicated from assets 1 and 2.

(d) Since $\theta^{1,*}$, $\theta^{2,*}$ and q^* is an equilibrium, the asset prices q^* are arbitrage-free. Agents have weakly monotonous utilities, hence by the FTAP there exist nonnegative state prices $\pi \geq 0$, such that $q = \pi' A$:

$$\begin{pmatrix} 1 \\ 4 \end{pmatrix} = (\pi_1, \pi_2, \pi_3) \begin{pmatrix} 1 & 0 \\ 0 & 1 \\ 0 & 1 \end{pmatrix}, \qquad \pi_1, \pi_2, \pi_3 > 0.$$

Solving with respect to π, we obtain: $\pi_1 = 1$ and $\pi_3 = 4 - \pi_2$ with $\pi_2 \in (0, 4)$. For a no arbitrage price of asset 3 we then have:

$$q^3 = \pi' A^3 = \pi_3,$$

or as $\pi_3 \in (0, 4)$, $q^3 \in (0, 4)$ is not uniquely determined.

(e) The financial market including the third asset is complete, hence any allocation of endowments is feasible and the market equilibrium can be determined via state prices:

$$\max_{c^1} \ln(c_1^1) + \ln(c_2^1), \quad \text{s.t.} \quad \begin{cases} \pi_1 c_1^1 + \pi_2 c_2^1 + \pi_3 c_3^1 = \pi_1 \cdot 0 + \pi_2 \cdot 1 + \pi_3 \cdot 2, \\ c_1^1 \geq 0, \quad c_2^1 \geq 0, \quad c_3^1 \geq 0, \end{cases} \tag{4.22}$$

$$\max_{c^2} \ln(c_2^2) + \ln(c_3^2), \quad \text{s.t.} \quad \begin{cases} \pi_1 c_1^2 + \pi_2 c_2^2 + \pi_3 c_3^2 = \pi_1 \cdot 2 + \pi_2 \cdot 1 + \pi_3 \cdot 0, \\ c_1^2 \geq 0, \quad c_2^2 \geq 0, \quad c_3^2 \geq 0, \end{cases} \tag{4.23}$$

$$c^1 + c^2 = w^1 + w^2 = 2. \tag{4.24}$$

Consider the optimization problem (4.22) of agent 1. The Lagrangian function is:

$$\mathcal{L}_1(c^1) = \ln(c_1^1) + \ln(c_2^1) - \lambda^1(\pi_1 c_1^1 + \pi_2 c_2^1 + \pi_3 c_3^1 - \pi_2 - 2\pi_3),$$

and the first order conditions are:

$$\frac{\partial \mathcal{L}_1}{\partial c_1^1} = \frac{1}{c_1^1} - \lambda^1 \pi_1 = 0,$$

$$\frac{\partial \mathcal{L}_1}{\partial c_2^1} = \frac{1}{c_2^1} - \lambda^1 \pi_2 = 0,$$

which after excluding λ^1 result in:

$$\frac{c_2^1}{c_1^1} = \frac{\pi_1}{\pi_2}.$$

Since consumption c_3^1 in state 3 has no influence on the agent 1's utility, it is chosen as small as possible, i.e. $c_3^1 = 0$.

Similarly, from the optimization problem (4.23) of agent 2 we obtain that $c_1^2 = 0$ and

$$\frac{c_3^2}{c_2^2} = \frac{\pi_2}{\pi_3}.$$

Further, by the market clearing in states 1 and 3 we have:

$$2 = c_1^1 + c_1^2 = c_1^1 + 0 \quad \text{and } 2 = c_3^1 + c_3^2 = 0 + c_3^2.$$

therefore, $c_1^1 = 2$, $c_3^2 = 2$,

$$c_2^1 = c_1^1 \frac{\pi_1}{\pi_2} = 2\frac{\pi_1}{\pi_2} \text{ and } c_2^2 = c_3^2 \frac{\pi_3}{\pi_2} = 2\frac{\pi_3}{\pi_2}.$$

Plugging the consumption into budget constraints of agents, we obtain:

$$\begin{aligned} &\text{agent 1:} && 2\pi_1 + 2\pi_1 = \pi_2 + 2\pi_3, \\ &\text{agent 2:} && 2\pi_3 + 2\pi_3 = 2\pi_1 + \pi_2, \end{aligned}$$

or equivalently, $\pi_1 = \pi_3$ and $\pi_2 = 2\pi_3$. Hence, for the optimal consumption we have:

$$\begin{aligned} c_1^1 &= 2, \quad c_2^1 = 1, \quad c_3^1 = 0, \\ c_1^2 &= 0, \quad c_2^2 = 1, \quad c_3^2 = 2. \end{aligned}$$

Recall, that by Walras Law, the market clearing in state 2, $2 = c_2^1 + c_2^2$, is satisfied too.

Next, for the asset prices q we obtain:

$$q = \pi' A = (\pi_1, \pi_2 + \pi_3, \pi_3)' = \pi_3(1, 3, 1)',$$

with

$$\frac{q_1}{q_2} = \frac{1}{3} \text{ and } \frac{q_2}{q_3} = 3.$$

Finally, the optimal portfolios can be found from the second period budget constraints: $c^i = w^i + A\theta^i$, $i = 1, 2$. Therefore, $\theta^i = A^-1(c^i - w^i)$ with

$$\theta^1 = (2, 0, -2) \text{ and } \theta^2 = (-2, 0, 2).$$

(f) In equilibrium the representative investors solves his utility maximization problem:

$$\max_{\theta} \frac{1}{5}\ln(2+\theta_1)+\frac{1}{5}\ln(2+\theta_2)+\frac{3}{5}\ln(2+\theta_2), \qquad \text{s.t. } q_1\theta_1+q_2\theta_2 \le 0,$$

and markets clear:

$$\theta_k = 0, \quad k = 1, 2.$$

The Lagrange function is:

$$\mathcal{L}(\theta) = \frac{1}{5}\ln(2+\theta_1)+\frac{1}{5}\ln(2+\theta_2)+\frac{3}{5}\ln(2+\theta_2)-\lambda(q_1\theta_1+q_2\theta_2),$$

and the first order conditions are

$$\frac{\partial \mathcal{L}}{\partial \theta_1} = \frac{1}{5(2+\theta_1)} - \lambda q_1 = 0,$$
$$\frac{\partial \mathcal{L}}{\partial \theta_2} = \frac{1}{5(2+\theta_2)} + \frac{3}{5(2+\theta_2)} - \lambda q_2 = 0.$$

Excluding λ and plugging $\theta_1 = \theta_2 = 0$ (market clearing), we obtain:

$$\frac{q_1}{q_2} = \frac{1}{4},$$

concluding the proof.

(g) The new decision problem of the representative investor is:

$$\max_{\theta} \frac{1}{5}\ln(2+\theta_1)+\frac{1}{5}\ln(2+\theta_2)+\frac{3}{5}\ln(2+\theta_2+\theta_3), \qquad \text{s.t. } q_1\theta_1+q_2\theta_2+q_3\theta_3 \le 0.$$

The corresponding Lagrange function and the FOCs are:

$$\mathcal{L}(\theta) = \frac{1}{5}\ln(2+\theta_1)+\frac{1}{5}\ln(2+\theta_2)+\frac{3}{5}\ln(2+\theta_2+\theta_3)-\lambda(q_1\theta_1+q_2\theta_2+q_3\theta_3),$$
$$\frac{\partial \mathcal{L}}{\partial \theta_1} = \frac{1}{5(2+\theta_1)} - \lambda q_1 = 0,$$
$$\frac{\partial \mathcal{L}}{\partial \theta_2} = \frac{1}{5(2+\theta_2)} + \frac{3}{5(2+\theta_2+\theta_3)} - \lambda q_2 = 0,$$
$$\frac{\partial \mathcal{L}}{\partial \theta_3} = \frac{3}{5(2+\theta_2+\theta_3)} - \lambda q_3 = 0.$$

From the market clearing for assets 1 and 2 we have $\theta_1 = \theta_2 = 0$, therefore, excluding λ, we obtain:

$$\frac{q_1}{q_2} = \frac{1}{4} \text{ and } \frac{q_2}{q_3} = \frac{4}{3}$$

in contrast to $\frac{q_1}{q_2} = \frac{1}{3}$ and $\frac{q_2}{q_3} = 3$ in the part (e). In other words, if the market structure is changed but the endowments remain constant, the representative agent stops doing his job.

5 Multiple-Periods Model

5.1 Financial market is **complete** if any consumption stream can be attained with at least one initial wealth. The necessary and sufficient condition for a financial market to be complete is that each two-period submarket is complete.

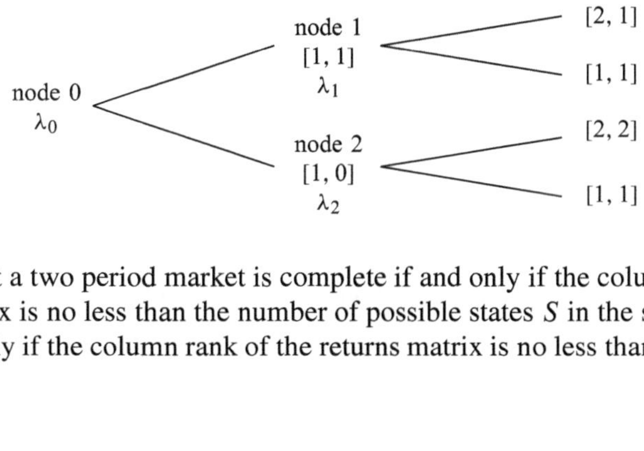

Recall that a two period market is complete if and only if the column rank of its payoffs matrix is no less than the number of possible states S in the second period, or equivalently if the column rank of the returns matrix is no less than S.

(i)

$$R_0 = \begin{pmatrix} 1 & 1 \\ 1 & 0 \end{pmatrix},$$

col_rk$(R_0) = 2 \geq S_0$, therefore the submarket in the node 0 is complete.

(ii)

$$R_1 = \begin{pmatrix} 2 & 1 \\ 1 & 1 \end{pmatrix},$$

col_rk$(R_1) = 2 \geq S_1$, therefore the submarket in the node 1 is complete.

T. Hens, M. O. Rieger, *Solutions to* Financial Economics,
Springer Texts in Business and Economics,
https://doi.org/10.1007/978-3-662-59889-4_13

(iii)

$$R_2 = \begin{pmatrix} 2 & 2 \\ 1 & 1 \end{pmatrix},$$

$\text{col\_rk}(R_2) = 1 < S_2$, therefore the submarket in the node 2 is incomplete.

Thereby, as there exists an incomplete two-period submarket, the total market is incomplete as well.

5.2 *Step 1.* A perfect foresight equilibrium in the model above is an *equilibrium with T-period fund separation*, that is for all $i = 1, \ldots, I$ and $t = 1, \ldots, T$,

$$\lambda_t^{i,k} = (1 - \lambda_t^{i,c})\bar{\lambda}_t^k, \quad k = 1, \ldots, K. \tag{5.1}$$

Consider the wealth dynamics of investor i. At time 0 he is given initial endowment w_0^i. At any further $t = 1, \ldots, T$:

$$w_t^i = \sum_{k=1}^{K} x_{t-1}^{i,k}(D_t^k + q_t^k) = \sum_{k=1}^{K} \frac{\lambda_{t-1}^{i,k} w_{t-1}^i}{q_{t-1}^k}(D_t^k + q_t^k) = w_{t-1}^i \sum_{k=1}^{K} \frac{(D_t^k + q_t^k)}{q_{t-1}^k} \lambda_{t-1}^{i,k}, \tag{5.2}$$

where $x_{t,k}^i$ is the amount of asset k investor i holds at time t.

Introduce portfolio returns $g_t^i = \sum_{k=1}^{K} \frac{(D_t^k + q_t^k)}{q_{t-1}^k} \lambda_{t-1}^{i,k}$, $t = 1, \ldots, T$. Then at any time t the wealth of the investor may be represented as $w_t^i = \prod_{\tau=1}^{t} g_\tau^i w_0^i$.

When choosing an investment strategy λ^i, each investor solves the following decision problem:

$$\begin{aligned}
\max_{\lambda^i} U^i(c^i) &= \mathbb{E}_P \sum_{t=1}^{T} (\delta^i)^t \log(c_t^i) \\
&= \mathbb{E}_P \sum_{t=1}^{T} (\delta^i)^t \ln(w_0^i \lambda_t^{i,c}) + \sum_{t=1}^{T} \left(\sum_{\tau=1}^{t} (\delta^i)^\tau \right) \mathbb{E}_P(g_t^i),
\end{aligned}$$

$$\text{s.t.} \begin{cases} c_t^i = \lambda_t^{i,c} w_t^i, \\ \lambda_t^{i,c} + \sum_{k=1}^{K} \lambda_t^{i,k} = 1, \\ w_t^i = \prod_{\tau=1}^{t} g_\tau^i w_0^i, \quad t = 1, \ldots, T. \end{cases} \tag{5.3}$$

We notice that maximization in (5.3) can be realized in two steps. First, for a given $\lambda_t^{i,c}$, $U^i(c^i)$ should be maximized over all $\lambda_t^{i,k}$, $k = 1, \ldots, K$, such that $\sum_{k=1}^{K} \lambda_t^{i,k} = 1 - \lambda_t^{i,c}$. Second, the obtained result should be maximized over all possible $\lambda_t^{i,c}$.

When consumption plan $\lambda_t^{i,c}$ is given, the decision problem (5.3) is equivalent to

$$\max_{\lambda^i} \sum_{t=1}^{T} \left(\sum_{\tau=1}^{t} (\delta^i)^\tau \right) \mathbb{E}_P(g_t^i), \quad \text{s.t.} \sum_{k=1}^{K} \lambda_t^{i,k} = 1 - \lambda_t^{i,c}, \quad t = 1, \ldots, T. \tag{5.4}$$

Because g_t^i are iid (they depend only on dividends process which is iid), (5.4) can be solved separately for each time t, that is

$$\max_{\lambda_t^i} \mathbb{E}_P(g_t^i), \quad \text{s.t.} \sum_{k=1}^{K} \lambda_t^{i,k} = 1 - \lambda_t^{i,c}.$$

After introducing $\bar{\lambda}_t^{i,k} = \frac{\lambda_t^{i,k}}{1-\lambda_t^{i,c}}$, the last can be rewritten as

$$\max_{\bar{\lambda}_t^i} \mathbb{E}_P \sum_{k=1}^{K} \frac{(D_t^k + q_t^k)}{q_{t-1}^k} \bar{\lambda}_{t-1}^{i,k}, \quad \text{s.t.} \sum_{k=1}^{K} \bar{\lambda}_{t-1}^{i,k} = 1, \tag{5.5}$$

where we took into account that $\lambda_t^{i,c}$ is given, and therefore does not influence the solution.

Finally, we notice that the objective function and the constraint in (5.5) do not depend on i. Therefore, its solution λ_t^k, $k = 1, \ldots, K$, does not depend on i. Hereby, for every i and a given consumption plan $\lambda_t^{c,i}$, the optimal investment strategy $\lambda_t^{i,k}$, $k = 1, \ldots, K$, satisfies (5.1), or equivalently, T-fund separation holds.

Step 2. See the proof of the Theorem 1 from "Dynamic General Equilibrium and T-Period Fund Separation", Gerber, Hens and Woehrmann, Journal of Financial and Quantitative Analysis, 2010.

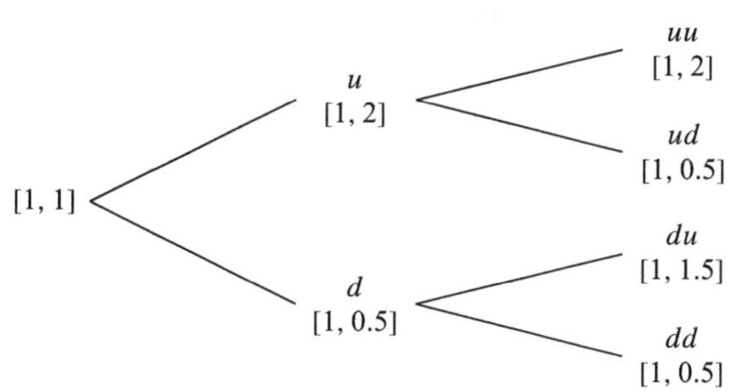

5.3

(a) The asset price (both bond and stock) in the second period, $t = 2$, is zero. This is the case, since the economy ends in $t = 2$, and therefore a stock bought at $t = 2$ does not pay out anything anymore.

(b) Denote the bond and the stock by $k = 0$ and $k = 1$ respectively. The representative agent has an expected log-utility, i.e.

$$U(c) = \mathbb{E}_P \sum_{t=0}^{2} \log(c_t) = \log(c_0) + p_u \log(c_{1,u}) + p_d \log(c_{1,d})$$
$$+ p_{uu} \log(c_{2,uu}) + p_{ud} \log(c_{2,ud}) + p_{du} \log(c_{2,du}) + p_{dd} \log(c_{2,dd}).$$

The decision problem of the representative investor is written as follows:

$$\max_\theta U(c), \quad \text{s.t.} \begin{cases} c_0 = \sum_{k=0,1}(D_0^k + q_0^k)\theta_{-1}^k - \sum_{k=0,1} q_0^k \theta_0^k, \\ c_{1,\omega_1} = \sum_{k=0,1}(D_{1,\omega_1}^k + q_{1,\omega_1}^k)\theta_0^k - \sum_{k=0,1} q_{1,\omega_1}^k \theta_{1,\omega_1}^k, \ \omega_1 = u, d, \\ c_{2,\omega_1\omega_2} = \sum_{k=0,1} D_{2,\omega_1\omega_2}^k \theta_{1,\omega_1}^k, \ \omega_1, \omega_2 = u, d, \end{cases} \tag{5.6}$$

where $\theta = (\theta_0^0, \theta_0^1, \theta_1^0, \theta_1^1, \theta_2^0, \theta_2^1)'$ is the assets portfolio the representative investor holds, q_t^k and D_t^k is the price and the dividend of the asset k in period t respectively, index ω_t corresponds to the state of the market in period t. In the budget constraint of the second period we used that the economy terminates, therefore asset prices are zero and investor forms no new portfolio.

Plugging in consumptions from budget constraints into the utility function results in the unconstrained maximization problem with the following FOCs:

$$\frac{\partial U}{\partial \theta_0^k} = -\frac{q_0^k}{c_0} + p_u \frac{D_{1,u}^k + q_{1,u}^k}{c_{1,u}} + p_d \frac{D_{1,d}^k + q_{1,d}^k}{c_{1,d}} = 0,$$

$$\frac{\partial U}{\partial \theta_{1,u}^k} = -p_u \frac{q_{1,u}^k}{c_{1,u}} + p_{uu} \frac{D_{2,uu}^k}{c_{2,uu}} + p_{ud} \frac{D_{2,ud}^k}{c_{2,ud}} = 0,$$

$$\frac{\partial U}{\partial \theta_{1,d}^k} = -p_d \frac{q_{1,d}^k}{c_{1,d}} + p_{du} \frac{D_{2,du}^k}{c_{2,du}} + p_{dd} \frac{D_{2,dd}^k}{c_{2,dd}} = 0,$$

where $k = 0, 1$. Rewriting the first order conditions we get the following equations for the asset prices:

$$q_0^k = c_0 \left(p_u \frac{D_{1,u}^k + q_{1,u}^k}{c_{1,u}} + p_d \frac{D_{1,d}^k + q_{1,d}^k}{c_{1,d}} \right), \tag{5.7}$$

$$q_{1,u}^k = \frac{c_{1,u}}{p_u} \left(p_{uu} \frac{D_{2,uu}^k}{c_{2,uu}} + p_{ud} \frac{D_{2,ud}^k}{c_{2,ud}} \right), \tag{5.8}$$

$$q_{1,d}^k = \frac{c_{1,d}}{p_d} \left(p_{du} \frac{D_{2,du}^k}{c_{2,du}} + p_{dd} \frac{D_{2,dd}^k}{c_{2,dd}} \right). \tag{5.9}$$

The market clearing condition with one representative agent is that $\theta_{t,s}^k = 1$ for both assets $k = 0, 1$ in all periods $t = 0, 1, 2$ and states ω_t. Plugging this into the budget constraints for $t = 0, 1$, we obtain that

$$\begin{aligned}
c_{t,\omega_t} &= \sum_{k=0,1} (D_{t,\omega_t}^k + q_{t,\omega_t}^k)\theta_{t-1,\omega_{t-1}}^k - \sum_{k=0,1} q_{t,\omega_t}^k \theta_{t,\omega_t}^k \\
&= \sum_{k=0,1} (D_{t,\omega_t}^k + q_{t,\omega_t}^k) \cdot 1 - \sum_{k=0,1} q_{t,\omega_t}^k \cdot 1 = \sum_{k=0,1} D_{t,\omega_t}^k, \quad t = 0, 1, \\
c_{2,\omega_1,\omega_2} &= \sum_{k=0,1} D_{2,\omega_1,\omega_2}^k.
\end{aligned}$$

The above means that in each (t, ω_t) the representative agent consumes total dividends, therefore

$$\begin{aligned}
&c_0 = 2, \\
&c_{1,u} = 3, \quad c_{1,d} = 1.5, \\
&c_{2,uu} = 3, \quad c_{2,ud} = 1.5, \quad c_{2,du} = 2.5, \quad c_{2,dd} = 1.5.
\end{aligned}$$

Finally, we recall that probabilities of different states are: $p_{\omega_1} = 1/2$ and $p_{\omega_1 \omega_2} = p_{\omega_1} \cdot p_{\omega_2} = 1/4$ for each $\omega_1, \omega_2 = u, d$. Hence, plugging all ingredients into the assets' prices equations (5.7)–(5.9), we find that

$$\begin{aligned}
q_0^0 &= \frac{61}{30}, \quad q_0^1 = \frac{59}{30}, \\
q_{1,u}^0 &= \frac{3}{2}, \quad q_{1,u}^1 = \frac{3}{2}, \\
q_{1,d}^0 &= \frac{4}{5}, \quad q_{1,d}^1 = \frac{7}{10}.
\end{aligned}$$

(c) Yes, the market is complete, since the payoff matrix of any two-period submarket is complete.

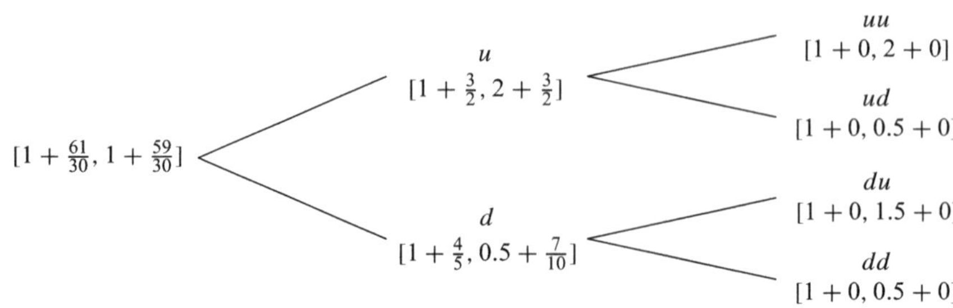

Indeed, the payoff matrix of the submarket in the period $t = 0$ is represented by the payoffs (price plus dividend) of assets in the following period $t = 1$, i.e.

$$A_0 = \begin{pmatrix} q^0_{1,u} + D^0_{1,u} & q^1_{1,u} + D^1_{1,u} \\ q^0_{1,d} + D^0_{1,d} & q^1_{1,d} + D^1_{1,d} \end{pmatrix} = \begin{pmatrix} 2.5 & 3.5 \\ 1.8 & 1.2 \end{pmatrix}.$$

$det(A_0) = -3.3 \neq 0$, therefore the first period submarket is complete.

Similarly we can check that both second-period submarkets, $t = 1$, are complete:

$$A_{1,u} = \begin{pmatrix} 1 & 2 \\ 1 & 0.5 \end{pmatrix}, \quad det(A_{1,u}) = -1.5 \neq 0,$$

$$A_{1,u} = \begin{pmatrix} 1 & 1.5 \\ 1 & 0.5 \end{pmatrix}, \quad det(A_{1,u}) = -1 \neq 0.$$

(d) Due to completeness of the market, as in the two-period model, the state prices π_{t,ω_t} can be seen as the prices in the following equilibrium model with one representative agent:

$$\max_c U(c) = \sum_{t=0}^{2} \sum_{\omega_t \in \Omega_t} p_{t,\omega_t} c_{t,\omega_t},$$

$$\text{s.t.} \sum_{t=0}^{2} \sum_{\omega_t \in \Omega_t} \pi_{t,\omega_t} c_{t,\omega_t} = \sum_{t=0}^{2} \sum_{\omega_t \in \Omega_t} \pi_{t,\omega_t} w_{t,\omega_t}, \tag{5.10}$$

with the market clearing condition

$$c_{t,\omega_t} = w_{t,\omega_t}, \quad t = 0, 1, 2, \quad \omega_t \in \Omega_t, \tag{5.11}$$

where Ω_t is the set of all possible states ω_t at time t, and $w_{t,\omega_t} = \sum_{k=0,1} D_{t,\omega_t}\theta_{-1}^k$ is an exogenous wealth endowment the representative investor receives in (t,ω_t). The state price π_{t,ω_t} may be seen as the price of an elementary security paying 1 in state ω_t of period t. We normalize $\pi_0 = 1$.

The budget constraint in (5.10) may be easily derived from the initial problem (5.6) as follows. First, we notice that for every asset k and time t,

$$\pi_{t,\omega_t} q_t^k = \sum_{\omega_{t+1}\in\Omega_{t+1}} \pi_{t+1,\omega_{t+1}}(q_{t+1,\omega_{t+1}}^k(\omega_t) + D_{t+1,\omega_{t+1}}^k(\omega_t)). \tag{5.12}$$

Second, we rewrite the first-period budget constraint in (5.6) using (5.12) and next periods constraints from (5.6):

$$\begin{aligned}
c_0 &= \sum_k D_0^k\theta_{-1}^k + \sum_k q_0^k(\theta_{-1}^k - \theta_0^k)\\
&= \sum_k D_0^k\theta_{-1}^k + \sum_{\omega_1\in\Omega_1}\pi_{1,\omega_1}\sum_k (q_{1,\omega_1}^k + D_{1,\omega_1}^k)(\theta_{-1}^k - \theta_0^k)\\
&= w_0 + \sum_{\omega_1\in\Omega_1}\pi_{1,\omega_1}\sum_k q_{1,\omega_1}^k\theta_{-1}^k + \sum_{\omega_1\in\Omega_1}\pi_{1,\omega_1}\sum_k D_{1,\omega_1}^k\theta_{-1}^k\\
&- \sum_{\omega_1\in\Omega_1}\pi_{1,\omega_1}\sum_k (q_{1,\omega_1}^k + D_{1,\omega_1}^k)\theta_0^k\\
&= w_0 + \sum_{\omega_1\in\Omega_1}\pi_{1,\omega_1}\sum_k q_{1,\omega_1}^k\theta_{-1}^k + \sum_{\omega_1\in\Omega_1}\pi_{1,\omega_1}w_{1,\omega_1}\\
&- \sum_{\omega_1\in\Omega_1}\pi_{1,\omega_1}\left(c_{1,\omega_1} + \sum_k q_{1,\omega_1}^k\theta_{1,\omega_1}^k\right)\\
&= w_0 + \sum_{\omega_2\in\Omega_2}\pi_{2,\omega_2}\sum_k D_{2,\omega_2}^k\theta_{-1}^k + \sum_{\omega_1\in\Omega_1}\pi_{1,\omega_1}w_{1,\omega_1}\\
&- \sum_{\omega_1\in\Omega_1}\pi_{1,\omega_1}c_{1,\omega_1} - \sum_{\omega_2\in\Omega_2}\pi_{2,\omega_2}\sum_k D_{2,\omega_2}^k\theta_{1,\omega_1}^k\\
&= w_0 + \sum_{\omega_2\in\Omega_2}\pi_{2,\omega_2}w_{2,\omega_2} + \sum_{\omega_1\in\Omega_1}\pi_{1,\omega_1}w_{1,\omega_1}\\
&- \sum_{\omega_1\in\Omega_1}\pi_{1,\omega_1}c_{1,\omega_1} - \sum_{\omega_2\in\Omega_2}\pi_{2,\omega_2}c_{2,\omega_2}.
\end{aligned}$$

The Lagrange function of the decision problem (5.10) is

$$\mathcal{L}(c) = \sum_{t=0}^{2}\sum_{\omega_t\in\Omega_t} p_{t,\omega_t}c_{t,\omega_t} - \lambda\sum_{t=0}^{2}\sum_{\omega_t\in\Omega_t}\pi_{t,\omega_t}(c_{t,\omega_t} - w_{t,\omega_t}),$$

and the corresponding FOCs are

$$\frac{\partial \mathcal{L}}{\partial c_{t,\omega_t}} = p_{t,\omega_t}\frac{1}{c_{t,\omega_t}} - \lambda\pi_{t,\omega_t} = 0.$$

Taking $\lambda\pi_{t,\omega_t}$ on the other side and dividing by the FOC of $t = 0$ results in

$$\frac{p_{t,\omega_t}c_0}{p_0 c_{t,\omega_t}} = \frac{\pi_{t,\omega_t}}{\pi_0},$$

or equivalently,

$$\pi_{t,\omega_t} = \frac{\pi_0}{p_0}p_{t,\omega_t}\frac{c_0}{c_{t,\omega_t}} = p_{t,\omega_t}\frac{c_0}{c_{t,\omega_t}}, \tag{5.13}$$

where we used that $\pi_0 = p_0 = 1$. The optimal consumption plan is known from the market clearing condition (5.11). Hereby, plugging the numbers into (5.13) results in

$$\begin{aligned}
&\pi_0 = 1,\\
&\pi_{1,u} = \frac{1}{3}, \quad \pi_{1,d} = \frac{2}{3},\\
&\pi_{2,uu} = \frac{1}{6}, \quad \pi_{2,ud} = \frac{1}{3}, \quad \pi_{2,du} = \frac{1}{5}, \quad \pi_{2,dd} = \frac{1}{3}.
\end{aligned}$$

(e) In part (d) we have shown that there are unique (up to multiplier π_0) strictly positive state prices in the market. By the FTAP this implies no arbitrage.
(f) The value of an European call is

$$\begin{aligned}
C_0^E &= \sum_{\omega_2\in\Omega_2} \pi_{2,\omega_2}(D^1_{2,\omega_2} - K)^+\\
&= \pi_{2,uu}(D^1_{2,uu}-K)^+ + \pi_{2,up}(D^1_{2,up}-K)^+ + \pi_{2,du}(D^1_{2,du}-K)^+ + \pi_{2,dd}(D^1_{2,dd}-K)^+\\
&= \frac{1}{6}\cdot 1 + \frac{1}{3}\cdot 0 + \frac{1}{5}\cdot\frac{1}{2} + \frac{1}{3}\cdot 0 = \frac{4}{15}.
\end{aligned}$$

(g) Unlike the European option, the American option can be exercised at any point in time. It needs to be decided at every point in time (and in every state), if it is more valuable to exercise the option or to wait and then exercise the option in later time. In the following the value of exercising the option and the value of waiting, both discounted to $t = 0$, are calculated.

Node $(1, u)$:

- Exercising:

$$\pi_{1,u}\left(q_{1,u}^{1}+D_{1,u}^{1}-K\right)^{+}=\frac{1}{3}\left(\frac{3}{2}+2-1\right)^{+}=\frac{5}{6}.$$

- Waiting:

$$\begin{aligned}\pi_{2,uu}\left(q_{2,uu}^{1}+D_{2,uu}^{1}-K\right)^{+}&+\pi_{2,ud}\left(q_{2,ud}^{1}+D_{2,ud}^{1}-K\right)^{+}\\&=\frac{1}{6}(0+2-1)^{+}+\frac{1}{3}(0+0.5-1)^{+}=\frac{1}{6}.\end{aligned}$$

The investor will therefore prefer to exercise the option. The discounted to $t=0$ value of the option for the node $(1, u)$ is

$$C_{1,u}^{A}=\max\left(\frac{5}{6},\frac{1}{6}\right)=\frac{5}{6}.$$

Node $(1, d)$:

- Exercising:

$$\pi_{1,d}\left(q_{1,d}^{1}+D_{1,d}^{1}-K\right)^{+}=\frac{2}{3}\left(\frac{7}{10}+0.5-1\right)^{+}=\frac{2}{15}.$$

- Waiting:

$$\begin{aligned}\pi_{2,du}\left(q_{2,du}^{1}+D_{2,du}^{1}-K\right)^{+}&+\pi_{2,dd}\left(q_{2,dd}^{1}+D_{2,dd}^{1}-K\right)^{+}\\&=\frac{1}{5}(0+1.5-1)^{+}+\frac{1}{3}(0+0.5-1)^{+}=\frac{1}{10}.\end{aligned}$$

The discounted to $t=0$ value of the option for the node $(1, d)$ is

$$C_{1,d}^{A}=\max\left(\frac{2}{15},\frac{1}{10}\right)=\frac{2}{15}.$$

Therefore, exercising the option turns out to be more valuable.

Node $t = 0$:

- Exercising:

$$\pi_0 \left(q_0^1 + D_0^1 - K\right)^+ = 1\left(\frac{59}{30} + 1 - 1\right)^+ = \frac{59}{30}.$$

- Waiting:

$$C_{1,u}^A + C_{1,d}^A = \frac{5}{6} + \frac{2}{15} = \frac{29}{30}.$$

The discounted to $t = 0$ value of the option for the node 0 is

$$C_0^A = \max\left(\frac{59}{30}, \frac{29}{30}\right) = \frac{29}{30}.$$

The option to exercise at $t = 0$ is more valuable than to wait. The value of the American option at $t = 0$ is $C_0^A = \frac{29}{30}$.

(h) The risk-free asset with maturity t costs 1 at time $t = 0$ and delivers a sure payoff $(1 + r_{f,t})^t$ at maturity t. Since the market is arbitrage free,

$$1 = \sum_{\omega_t \in \Omega_t} \pi_{t,\omega_t}(1 + r_{f,t})^t = (1 + r_{f,t})^t \sum_{\omega_t \in \Omega_t} \pi_{t,\omega_t}.$$

Therefore,

$$R_{f,t} = 1 + r_{f,t} = \left(\frac{1}{\sum_{\omega_t \in \Omega_t} \pi_{t,\omega_t}}\right)^{1/t},$$

$$R_{f,1} = \frac{1}{\pi_{1,u} + \pi_{1,d}} = 1,$$

$$R_{f,2} = \left(\frac{1}{\pi_{2,uu} + \pi_{2,ud} + \pi_{2,du} + \pi_{2,dd}}\right)^{1/2} = \sqrt{\frac{30}{31}}.$$

(i) No arbitrage implies automatically the results.

(i) $\pi_0^* = \frac{\pi_1}{\pi_{1,u} + \pi_{1,d}} = \left(\frac{1}{3}, \frac{2}{3}\right)'$.

$$\frac{1}{1 + r_{f,1}} \mathbb{E}_{\pi_0^*}(D_1^0 + q_1^0) = 1 \cdot \frac{1}{3} \cdot \left(1 + \frac{3}{2}\right) + 1 \cdot \frac{2}{3} \cdot \left(1 + \frac{4}{5}\right) = \frac{61}{30} = q_0^0,$$

$$\frac{1}{1 + r_{f,1}} \mathbb{E}_{\pi_0^*}(D_1^1 + q_1^1) = 1 \cdot \frac{1}{3} \cdot \left(2 + \frac{3}{2}\right) + 1 \cdot \frac{2}{3} \cdot \left(\frac{1}{2} + \frac{7}{10}\right) = \frac{59}{30} = q_0^1.$$

(ii)

$$R_{t+1}^k = \frac{D_{t+1}^k + q_{t+1}^k}{q_t^k},$$

$$R_{1,u}^0 = \frac{1+3/2}{61/30} = \frac{75}{61}, \quad R_{1,d}^0 = \frac{1+4/5}{61/30} = \frac{54}{61},$$

$$R_{1,u}^1 = \frac{2+3/2}{59/30} = \frac{105}{59}, \quad R_{1,d}^1 = \frac{1/2+7/10}{59/30} = \frac{36}{59}.$$

$$\mathbb{E}_{\pi_0^*}(R_1^0) = \pi_{0,u}^* R_{1,u}^0 + \pi_{0,d}^* R_{1,d}^0 = \frac{1}{3}\cdot\frac{75}{61} + \frac{2}{3}\cdot\frac{54}{61} = 1 = R_{f,1},$$

$$\mathbb{E}_{\pi_0^*}(R_1^1) = \pi_{0,u}^* R_{1,u}^1 + \pi_{0,d}^* R_{1,d}^1 = \frac{1}{3}\cdot\frac{105}{59} + \frac{2}{3}\cdot\frac{36}{59} = 1 = R_{f,1}.$$

(iii) For the likelihood process we have $\ell_{1,u} = \frac{1/3}{1/2} = \frac{2}{3}$, $\ell_{1,d} = \frac{2/3}{1/2} = \frac{4}{3}$. Furthermore we get

$$\mathbb{E}_{P_0}(R_1^0) = \frac{1}{2}\cdot\frac{75}{61} + \frac{1}{2}\cdot\frac{54}{61} = \frac{129}{122},$$

$$\mathrm{cov}_{P_0}(R_1^0, \ell_1) = \mathbb{E}_{P_0}(R_1^0\ell_1) - \mathbb{E}_{P_0}(R_1^0)\mathbb{E}_{P_0}(\ell_1) = 1 - \frac{129}{2\cdot 61}\cdot 1 = -\frac{7}{122}.$$

Plug that in:

$$R_{f,1} - \mathrm{cov}_{P_0}(R_1^0, \ell_1) = 1 + \frac{7}{122} = \frac{129}{122} = \mathbb{E}_{P_0}(R_1^0).$$

Analogously we check for $k = 1$.

5.4 Because of no arbitrage the following relation must hold:

$$(1 + r_{t_0,t_1})^{t_1}\,(1 + f(t_0, t_1, t_2))^{t_2 - t_1} = (1 + r_{t_0,t_2})^{t_2},$$

where r_{t_0,t_1} is the annual interest rate of a bond which has in t_0 a maturity of $t_1 - t_0$ and $f(t_0, t_1, t_2)$ is the forward rate between t_1 and t_2 from a forward traded at t_0. For the forward rate we then obtain that

$$1 + f(t_0, t_1, t_2) = \frac{(1 + r_{t_0,t_2})^{\frac{t_2}{t_2 - t_1}}}{(1 + r_{t_0,t_1})^{\frac{t_1}{t_2 - t_1}}}.$$

This implies $f(0, 5, 8) = 8.4\%$.

5.5

(a) The utility maximization problem of the representative investor is:

$$\max_{c_0,c_1,c_2} \ln(c_0) + \frac{1}{1+\delta}\ln(c_1) + \frac{1}{(1+\delta)^2}\ln(c_2),$$

$$\text{s.t. } p_0c_0 + s_{01} + s_{02} = p_0w_0,$$

$$p_1c_1 + s_{12} = p_1w_1 + (1+r_{01})s_{01},$$

$$p_2c_2 = p_2w_2 + (1+f_{12})s_{12} + (1+r_{02})^2 s_{02},$$

where s_{01}, s_{02} and s_{12} are the investments into the bonds and the forward. The market clearing conditions are $c_0 = w_0$, $c_1 = w_1$ and $c_2 = w_2$. This problem can be solved analogously to Exercise 5.7. We get

$$1 + r_{01} = (1+\delta)(1+\overline{g}_{01}), \qquad 1 + r_{02} = (1+\delta)\sqrt{1+\overline{g}_{02}},$$

$$1 + f_{12} = (1+\delta)(1+\overline{g}_{12}),$$

where $\overline{g}_{t\ t+1}$ is the nominal growth rate between t and $t+1$. Till here we know the interest rates in $t = 0$. In $t = 1$ the economy is exactly the same as in Exercise 5.7. r_{12}, the interest rate realized in $t = 1$, is then:

$$1 + r_{12} = (1+\delta)(1+\overline{g}_{12}).$$

(b) The utility maximization problem of the representative investor is:

$$\max_{c_0,c_1,c_2} \ln(c_0) + \frac{1}{1+\beta}\left(\frac{1}{1+\delta}\ln(c_1) + \frac{1}{(1+\delta)^2}\ln(c_2)\right),$$

$$\text{s.t. } p_0c_0 + s_{01} + s_{02} = p_0w_0,$$

$$p_1c_1 + s_{12} = p_1w_1 + (1+r_{01})s_{01},$$

$$p_2c_2 = p_2w_2 + (1+f_{12})s_{12} + (1+r_{02})^2 s_{02}.$$

The market clearing conditions are $c_0 = w_0$, $c_1 = w_1$ and $c_2 = w_2$. This problem can be solved in the same way as in Exercise 5.7. We get

$$1 + r_{01} = (1+\beta)(1+\delta)(1+\overline{g}_{01}), \quad 1 + r_{02} = \sqrt{1+\beta}\,(1+\delta)\sqrt{1+\overline{g}_{02}},$$

$$1 + f_{12} = (1+\delta)(1+\overline{g}_{12}),$$

where $\overline{g}_{t\ t+1}$ is the nominal growth rate between t and $t+1$. Till here we know the interest rates in $t = 0$. In $t = 1$ the economy is exactly the same as in

Exercise 5.7. Therefore, r_{12}, the interest rate realized in $t = 1$, is

$$1 + r_{12} = (1 + \beta)(1 + \delta)(1 + \overline{g}_{12}).$$

The utility function to obtain that result was: $\ln(c_1) + \frac{1}{1+\beta}\frac{1}{1+\delta}\ln(c_2)$.

(c)

(i) The utility maximization problem of the representative investor is:

$$\max_{c_0, c_{1u}, c_{2u}, c_{1d}, c_{2d}} \ln(c_0) + q\left(\frac{1}{1+\delta}\ln(c_{1u}) + \frac{1}{(1+\delta)^2}\ln(c_{2u})\right)$$
$$+ (1-q)\left(\frac{1}{1+\delta}\ln(c_{1d}) + \frac{1}{(1+\delta)^2}\ln(c_{2d})\right),$$
$$\text{s.t. } p_0c_0 + s_{01} + s_{02} = p_0w_0,$$
$$p_{1u}c_{1u} + s_{12} = p_{1u}w_{1u} + (1 + r_{01})s_{01},$$
$$p_1c_{1d} + s_{12} = p_{1d}w_{1d} + (1 + r_{01})s_{01},$$
$$p_{2u}c_{2u} = p_{2u}w_{2u} + (1 + f_{12})s_{12} + (1 + r_{02})^2 s_{02},$$
$$p_{2d}c_{2d} = p_{2d}w_{2d} + (1 + f_{12})s_{12} + (1 + r_{02})^2 s_{02},$$

where s_{01}, s_{02} and s_{12} are the investments into the bonds and the forward. The market clearing conditions are $c_0 = w_0$, $c_1 = w_1$ and $c_2 = w_2$. This problem can be solved analogously to Exercise 5.7. We get

$$1 + r_{01} = \frac{1+\delta}{\mathbb{E}\left(\frac{1}{1+\overline{g}_{01}}\right)}, \qquad 1 + r_{02} = \frac{1+\delta}{\sqrt{\mathbb{E}\left(\frac{1}{1+\overline{g}_{02}}\right)}},$$

$$1 + f_{12} = (1+\delta)\frac{\mathbb{E}\left(\frac{1}{1+\overline{g}_{01}}\right)}{\mathbb{E}\left(\frac{1}{1+\overline{g}_{02}}\right)},$$

where $\overline{g}_{t\,t+1}$ is the nominal growth rate between t and $t+1$. Till here we know the interest rates in $t = 0$. In $t = 1$ the economy is exactly the same as in Exercise 5.7. Therefore, r_{12s}, the interest rate realized in $t = 1$, if the economy is in state s, is as follows:

$$1 + r_{12u} = (1+\delta)(1 + \overline{g}_{12u}), \qquad 1 + r_{12d} = (1+\delta)(1 + \overline{g}_{12d}),$$
$$\mathbb{E}(1 + r_{12}) = (1+\delta)\mathbb{E}(1 + \overline{g}_{12}),$$

where we considered the upper state and the lower state of $t = 1$ separately.

(ii) First some helpful calculations:

$$\mathbb{E}\left(\frac{1}{1+\overline{g}_{01}}\right) = 0.5\frac{9}{10} + 0.5\frac{21}{20} = \frac{39}{40},$$

$$\mathbb{E}\left(\frac{1}{1+\overline{g}_{02}}\right) = 0.5\left(\frac{9}{10}\right)^2 + 0.5 \cdot 1 = \frac{181}{200},$$

$$1+\overline{g}_{01d} = \frac{1+\overline{g}_{02d}}{1+\overline{g}_{01d}} = \frac{21}{20}.$$

For the different interest rates we then obtain:

$$1+r_{01} = 1.1 \cdot \frac{40}{39} = 1.128,$$

$$1+r_{02} = 1.1\sqrt{\frac{200}{181}} = 1.156,$$

$$1+f_{12} = 1.1 \cdot \frac{195}{181} = 1.185.$$

In $t = 1$ the realized interest rates depend on the state:

$$1+r_{12u} = 1.1 \cdot \frac{10}{9} = 1.222,$$

$$1+r_{12d} = 1.1 \cdot \frac{21}{20} = 1.155,$$

$$\mathbb{E}(1+r_{12}) = 1.1 \cdot \frac{389}{360} = 1.189.$$

(d) In the pure rational case without uncertainty an increasing or decreasing term structure can be obtained by choosing the right growth rates. But there is no forward rate bias (i.e. $f(0,1,2) = f_{12} = r_{12}$).

In the case with hyperbolic discounting, the term structure can also be explained. Plus there is a negative (but constant) forward rate bias (i.e. $f_{12} - r_{12} < 0$.

In the rational case with uncertainty, the shape of the term structure is also explained. The numerical example shows that in the upper state the realized interest rate rises and we have a negative forward rate bias. In the down state just the opposite happens. This is in line with the empirical observations. Since the model determines just one forward rate, it is not able to tell anything about the persistence of the forward rate bias. Empirical evidence shows that the forward rate bias has for several years the same sign.

5.6 λ^* invests according to the expected relative dividend. The matrix of relative dividends is:

$$\begin{pmatrix} \frac{3}{4} & \frac{1}{6} \\ \frac{1}{4} & \frac{5}{6} \end{pmatrix}$$

If we take column-wise expectations (for each asset) we find that λ^* invests 46% in asset 1 and 54% in asset 2.

5.7

(a) $\hat{\lambda}^M$ is a portfolio rule which invests proportional to the market capitalization,

$$\hat{\lambda}^M_{t,0} = \lambda_0,$$

$$\hat{\lambda}^M_{t,k} = \frac{p_t^k \cdot 1}{W_t} = \frac{\sum_{i=1}^I \lambda^i_{t,k} w^i_t}{W_t} = \sum_{i=1}^I \lambda^i_{t,k} r^i_t, \quad k = 1, \dots, K.$$

Therefore,

$$g(\hat{\lambda}, \hat{\lambda}^M)\,|_{\hat{\lambda}=\hat{\lambda}^M} = \mathbb{E}_P \ln\left((1-\lambda^c) + \lambda^c \sum_{k=1}^K \frac{d^k \hat{\lambda}^M_k}{\hat{\lambda}^M_k}\right)$$

$$= \mathbb{E}_P \ln\left((1-\lambda^c) + \lambda^c \sum_{k=1}^K d^k\right) = \mathbb{E}_P \ln(1) = 0.$$

(b) $\lambda^{1/K} = (1/K, \dots, 1/K)$ is the portfolio rule which invests equal proportions of its wealth into all long-lived assets. Hence,

$$g(\lambda^{1/K}, \hat{\lambda}) = \mathbb{E}_P \ln\left(1 - \lambda^c + \frac{1}{K}\lambda^c \sum_{k=1}^K d^k/\hat{\lambda}_k\right).$$

One can easily construct a portfolio rule $\hat{\lambda}$, such that $1-\lambda^c+K\lambda^c\sum_{k=1}^K d^k\hat{\lambda}_k >$ 1 almost surely. In this case $g(\hat{\lambda}, \hat{\lambda}^{1/K}) > 0$, therefore the claim of 1(b) is not correct. Since $\mathbb{E}_P(d^k) > 0$ some k as $\hat{\lambda}_k = \varepsilon \to 0$ $g(\lambda^{1/K}, \hat{\lambda}) > 0$

(c) Since logarithm is a concave function, the growth rate $g(\hat{\lambda}, \hat{\lambda}^*)$ is a concave function of $\hat{\lambda}$. Any local maximum of concave function is its global maximum. Therefore, to prove the assertion it is sufficient to show that $g(\hat{\lambda}, \lambda^*)$ has a local maximum at $\hat{\lambda} = \hat{\lambda}^*$, i.e.

$$\left.\frac{\partial g(\lambda^* + h\mu, \lambda^*)}{\partial h}\right|_{h=0} = 0$$

for any $\mu = (\mu_1, \ldots, \mu_K)$ such that $\sum_{k=1}^{K} \mu_k = 0$. Indeed,

$$
\begin{aligned}
\left.\frac{\partial g(\lambda^* + h\mu, \lambda^*)}{\partial h}\right|_{h=0} &= \sum_{k=1}^{K} \left.\frac{\partial g(\hat{\lambda}, \lambda^*)}{\partial \hat{\lambda}_k}\right|_{\hat{\lambda}=\lambda^*} \cdot \mu_k \\
&= \sum_{k=1}^{K} \frac{\partial}{\partial \hat{\lambda}_k} \mathbb{E}_P \ln\left(1 - \lambda_0 + \lambda_0 \sum_{l=1}^{K} \frac{d^l \hat{\lambda}_l}{\lambda_l^*}\right)\Bigg|_{\hat{\lambda}=\lambda^*} \cdot \mu_k \\
&= \sum_{k=1}^{K} \mathbb{E}_P \left(\frac{\lambda_0 \frac{d^k}{\lambda_k^*}}{1 - \lambda_0 + \lambda_0 \sum_{l=1}^{K} \frac{d^l \hat{\lambda}_l}{\lambda_l^*}}\right)\Bigg|_{\hat{\lambda}=\lambda^*} \cdot \mu_k \\
&= \sum_{k=1}^{K} \mathbb{E}_P \left(\frac{\lambda_0 \frac{d^k}{\mathbb{E}_P d^k}}{1 - \lambda_0 + \lambda_0 \sum_{l=1}^{K} d^l}\right) \cdot \mu_k \\
&= \sum_{k=1}^{K} \mathbb{E}_P \left(\lambda_0 \frac{d^k}{\mathbb{E}_P d^k}\right) \cdot \mu_k \\
&= \lambda_0 \sum_{k=1}^{K} \mu_k = 0.
\end{aligned}
$$

6 Theory of the Firm

6.1 $f = \{(y_0, y_1) \mid y_1 \leq \sqrt{-y_0},\ y_0 \leq 0\}$

(i) f is closed: defined by ≤ 0
(ii) f convex: since $\sqrt{\ }$ is concave
(iii) $y_0 < 0,\ y_1 > 0$
(iv) $y_0 = 0 \implies y_1 = 0$
(v) max production is $\sqrt{w_0}$

6.2

(a) 100% equity $\implies \xi^j = 0$
(b) 100% bonds
investment of $-y_0^j = 1$, needs to be financed on asset market, e.g. $\xi_2^j = 0$ and $\xi_1^j = \frac{-1}{1+r}$ where $\frac{1}{1+r} = q_1$. This implies the following dividends in period 1 ($d_0^j = 1$):

$$d_1^j = 1 - \frac{1}{1+r},\ d_2^j = 1.5 - \frac{1}{1+r}$$

(c) 100% risk coverage $\implies$ hedge production plan

$$\begin{pmatrix} 1 \\ -1.5 \end{pmatrix} = \begin{pmatrix} 1 & 1 \\ 1 & 0 \end{pmatrix} \begin{pmatrix} \xi_1^j \\ \xi_2^j \end{pmatrix} \qquad \implies \xi_1^j = -1.5$$

$$\implies -1 = -1.5 + \xi_2^j \qquad \implies \xi_2^j = 0.5$$

T. Hens, M. O. Rieger, *Solutions to* Financial Economics,
Springer Texts in Business and Economics,
https://doi.org/10.1007/978-3-662-59889-4_14

6.3 Using the formula from page 258 of the book:

$$\hat{z}^i = z^i + \frac{1}{100}\left[(\xi^1 - \hat{\xi}^1) + (\xi^2 - \hat{\xi}^2)\right]$$

$$\Longrightarrow \hat{z}^i = \begin{pmatrix}1\\1\end{pmatrix} + \frac{1}{100}\left[\begin{pmatrix}-1\\1\end{pmatrix} - \begin{pmatrix}1\\0\end{pmatrix} + \begin{pmatrix}1\\-1\end{pmatrix} - \begin{pmatrix}-1\\1\end{pmatrix}\right]$$

$$= \begin{pmatrix}1\\1\end{pmatrix} + \frac{1}{100}\begin{pmatrix}0\\-1\end{pmatrix} = \begin{pmatrix}1\\\frac{99}{100}\end{pmatrix}$$

6.4

(a) unlimited liability

$$d^j = y^j + \begin{pmatrix}-q'\\A\end{pmatrix}\xi^j,\ y^j = \begin{pmatrix}1\\2\end{pmatrix}$$

$$\langle A, D\rangle = \left\langle y^j\right\rangle + \langle A\rangle \text{ independent of } \xi^j$$

(b) limited liability

$$A = \begin{pmatrix}1\\1\end{pmatrix},\ d_s^j = \max\left\{0,\ y^j + \begin{pmatrix}-q\\A\end{pmatrix}\xi^j\right\}$$

$$\langle D\rangle \overset{\xi^j=0}{=} \left\langle\begin{pmatrix}1\\3\end{pmatrix}\right\rangle,\ \langle D\rangle \overset{\xi^j=-2}{=} \left\langle\begin{pmatrix}1\\3\end{pmatrix} - \begin{pmatrix}2\\2\end{pmatrix}\right\rangle^+ = \left\langle\begin{pmatrix}1\\1\end{pmatrix}\right\rangle$$

$$\text{Thus } \langle A, D\rangle \overset{\xi^j=0}{=} \left\langle\begin{pmatrix}1\\1\end{pmatrix}, \begin{pmatrix}1\\3\end{pmatrix}\right\rangle$$

$$\langle A, D\rangle \overset{\xi^j=-2}{=} \left\langle\begin{pmatrix}1\\1\end{pmatrix}\right\rangle$$

6.5

(a) Consumer 1's point of view:

$$\max_{x_0^1, x_1^1}\ ln(x_0^1) + 2ln(x_1^1) \quad s.t.\ x_1^1 = \sqrt{1 - x_0^1}$$

$$\max_{x_0^1}\ ln(x_0^1) + \underbrace{2ln(\sqrt{1 - x_0^1})}_{ln(1-x_0^1)}$$

$$\text{FOC} \quad \frac{1}{x_0^1} - \frac{1}{1-x_0^1} = 0$$

$$\Longrightarrow x_0^1 = \frac{1}{2},\ x_1^1 = \frac{1}{\sqrt{2}}$$

Consumer 2's point of view:

$$\max_{x_0^2} \ 2ln(x_0^2) + ln\sqrt{1-x_0^2}$$

$$\frac{2}{x_0^2} = \frac{1}{2}\frac{1}{1-x_0^2}$$

$$2(1-x_0^2) = \frac{1}{2}x_0^2$$

$$4 - 4x_0^2 = x_0^2$$

$$\Longrightarrow x_0^2 = \frac{4}{5},\ x_1^2 = \frac{1}{\sqrt{5}}$$

(b) The firm maximizes discounted intertemporal profits

$$\max \Pi = y_0 + \frac{1}{1+r}y_1, \quad s.t. \quad y_1 \le \sqrt{-y_0}$$

The consumers maximize their utilities

$$\max_{x_0^1,\, x_1^1} \ ln(x_0^1) + 2ln(x_1^1) \quad s.t. \quad x_0^1 + \frac{1}{1+r}x_1^1 \le 1 + \frac{1}{2}\Pi$$

$$\max_{x_0^2,\, x_1^2} \ 2ln(x_0^2) + ln(x_1^2) \quad s.t. \quad x_0^2 + \frac{1}{1+r}x_1^2 \le 1 + \frac{1}{2}\Pi$$

And markets clear

$$x_0^1 + x_0^2 = y_0 + 1 + 1$$

$$x_1^1 + x_1^2 = y_1$$

The first order condition for profit maximization is

$$1 - \frac{1}{1+r}\frac{1}{2\sqrt{-y_0}} = 0, \text{ which yields}$$

$$y_0 = -\frac{1}{4}\frac{1}{(1+r)^2} \text{ and } y_1 = \frac{1}{2}\frac{1}{1+r}$$

Moreover, the profits result in

$$\Pi = \frac{1}{4}\frac{1}{(1+r)^2}$$

Substituting $x_0^1 = 1+\frac{1}{2}\Pi - \frac{1}{1+r}x_1^1$ from the budget constraint, the maximization problem of the first household becomes

$$\max_{x_1^1}\ ln(1+\frac{1}{2}\Pi - \frac{1}{1+r}x_1^1) + 2ln(x_1^1).$$

The first order condition is

$$-\frac{1}{1+r}\left(\frac{1}{1+\frac{1}{2}\Pi - \frac{1}{1+r}x_1^1}\right) + \frac{2}{x_1^1} = 0, \text{ which yields}$$

$$x_1^1 = \frac{2+\Pi}{3}(1+r) \text{ and}$$

$$x_0^1 = \frac{1}{6}(2+\Pi).$$

Similarly, $x_0^2 = 1 + \frac{1}{2}\Pi - \frac{1}{1+r}x_1^2$, yields

$$\max_{x_1^2}\ 2ln(1+\frac{1}{2}\Pi - \frac{1}{1+r}x_1^2) + ln(x_1^2), \text{ which has the FOC}$$

$$\frac{-2}{1+\frac{1}{2}\Pi - \frac{1}{1+r}x_1^2}(1+r) + \frac{1}{x_1^2} = 0, \text{ resulting in}$$

$$x_1^2 = \frac{(1+r)(1+\frac{1}{2}\Pi)}{3} \text{ and}$$

$$x_0^2 = \frac{1}{3}(2+\Pi)$$

Market clearing for the first commodity requires

$$\frac{1}{6}(2+\Pi) + \frac{1}{3}(2+\Pi) = 2 - \frac{1}{4(1+r)^2}$$

(Note that by Walras' Law this condition is sufficient to clear both markets.) Solving for the equilibrium interest rate we obtain $(1+r) = \sqrt{\frac{3}{8}}$. Thus the equilibrium allocation is

$$y_0 = -\frac{1}{4}\frac{3}{8} = -\frac{2}{3}, \qquad y_1 = \frac{1}{2}\sqrt{\frac{8}{3}}, \quad \Pi = \frac{2}{3}$$

$$x_0^1 = \frac{1}{6}(2+\Pi) = \frac{4}{9}, \qquad x_1^1 = \frac{(2+\Pi)(1+r)}{3} = \frac{1}{3}\sqrt{\frac{8}{3}}$$

$$x_0^2 = \frac{1}{3}(2+\Pi) = \frac{8}{9}, \qquad x_1^2 = \frac{(1+\frac{1}{2}\Pi)(1+r)}{3} = \frac{1}{6}\sqrt{\frac{8}{3}}$$

Double checking market clearing:

$$\frac{4}{9} + \frac{8}{9} = 2 - \frac{2}{3},$$
$$\frac{12}{9} = \frac{4}{3} \text{ is okay,}$$

and

$$\frac{(2+\Pi+1+\frac{1}{2}\Pi)(1+r)}{3} = \frac{1}{2}\sqrt{\frac{8}{3}}$$
$$\frac{(3+\frac{3}{2}\Pi)(1+r)}{3} = \frac{1}{2}\sqrt{\frac{8}{3}}$$
$$(1+\frac{1}{2}\frac{2}{3})(1+r) = \frac{1}{2}\sqrt{\frac{8}{3}}$$
$$\frac{4}{3}\sqrt{\frac{3}{8}} = \frac{1}{2}\sqrt{\frac{8}{3}}$$
$$\frac{1}{2}\frac{\sqrt{8}\sqrt{8}\sqrt{3}}{3\sqrt{8}} = \frac{1}{2}\sqrt{\frac{8}{3}} \text{ is okay as well.}$$

6.6 In a Drèze equilibrium firms do not use market prices but it's owners' utility gradients to evaluate the firm's production plans. This is useful in particular when markets are not complete so that there is not a sufficient system of market prices. Let d denote the discount factor with which the firm evaluates production plans. Then its profit maximization problem is

$$\max_{0 \geq y_0,\ y_1 \geq 0} \Pi = \Pi_0 + d\Pi_1, \text{ where } \Pi_0 = Y_0 \text{ and } \Pi_1 = y_1 \quad \text{s.t.} \quad y_1 \leq \sqrt{-y_0}$$

With a capital market, as in 6.5(b), $d = \frac{1}{1+r}$. According to the Drèze criterion we determine d as follows

$$d = \frac{1}{2}\frac{U'^1(x_1^1)}{U'^1(x_0^1)} + \frac{1}{2}\frac{U'^2(x_1^2)}{U'^2(x_0^2)} \quad \text{(Recall both consumers hold 50\% of the shares.)}$$

The solution of the firm's profit maximization leads to $y_0 = -\frac{d^2}{4} = \Pi_0$, $y_1 = \frac{d}{2} = \Pi_1$. Recall the utility function of the consumers $U^1(x^1) = ln(x_0^1) + 2ln(x_1^1)$, $U^2(x^2) = 2ln(x_0^2) + ln(x_1^2)$. Then the discount factor d is given by

$$d = \frac{x_0^1}{x_1^1} + \frac{x_0^2}{4x_1^2}.$$

The budget restrictions are

$$x_0^1 = 1 + \frac{1}{2}\Pi_0 = \frac{8 - d^2}{8}, \qquad x_1^1 = \frac{1}{2}\Pi_1 = \frac{1}{4}d$$

$$x_0^2 = 1 + \frac{1}{2}\Pi_0 = \frac{8 - d^2}{8}, \qquad x_1^2 = \frac{1}{4}d$$

Inserting this into the formula for the discount factor yields:

$$d = \frac{\frac{8-d^2}{8}}{\frac{1}{4}d} + \frac{\frac{8-d^2}{16}}{\frac{1}{2}d} \iff \frac{4}{3}d^2 = 8 - d^2 \iff \frac{1}{3}d^2 = 8$$

$$d = \sqrt{\frac{8}{3}}$$

which coincides with the market interest factor $\frac{1}{1+r}$ from 6.5(b). The production plan of the firm is also as in 6.5(b). Finally the equilibrium allocation of consumption is

$$x_0^1 = \frac{8 - \frac{8}{3}}{8} = \frac{2}{3} = x_0^2, \quad x_1^1 = \frac{1}{4}\sqrt{\frac{8}{3}} = x_1^2$$

which is more symmetric than in the competitive equilibrium.

6.7

(a)

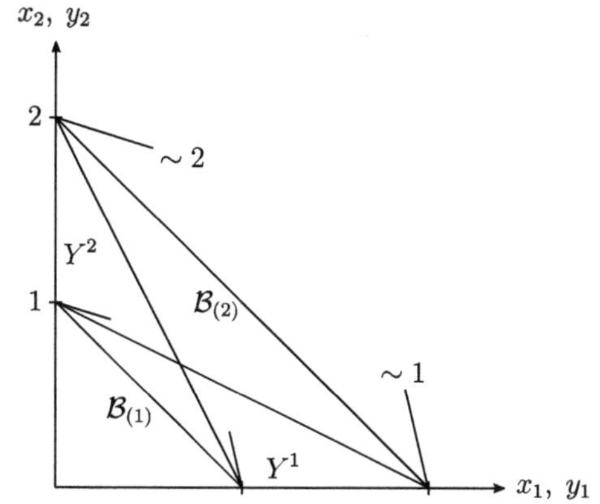

(b) For both equilibria the market clearing conditions, i.e. (iii)–(v) in the definition of the Drèze equilibrium for incorporated companies, are easily verified. Left to check are (i) utility maximization and (ii) profit maximization: 1. Drèze equilibrium:

(i) The budget set is

$$\mathcal{B}_{(1)} = \left\{ x \in \mathbb{R}^3_+ \mid \begin{pmatrix} x_0 \\ x_1 \\ x_2 \end{pmatrix} \leq \underbrace{\begin{pmatrix} 1 \\ 0 \\ 0 \end{pmatrix}}_{\omega} + \underbrace{\begin{pmatrix} -1 \\ 0 \\ 1 \end{pmatrix}}_{y^1} z_1 + \underbrace{\begin{pmatrix} -1 \\ 1 \\ 0 \end{pmatrix}}_{y^2} z_2 \right\}.$$

Since the consumer does not enjoy utility from x_0, but from x_1 and x_2, and since $y_1^k,\ y_2^k \geq 0,\ k = 1, 2$ he will fully invest his resources ω_o^i in the production of the firm. On the other hand, since $\omega_1^i = \omega_2^i = 0$, the consumer chooses from the set $\{x \in \mathbb{R}^2_+ \mid x_1 + x_2 \leq 1\}$ (see figure in (a)).

From this it follows that the figure in (a) already depicts the relevant decision. The consumers choose: $\overset{*}{x}{}^1 = (0, 1, 0)^T,\ \overset{*}{x}{}^2 = (0, 0, 1)^T,\ \overset{*}{z}{}^1 = (0, 1)^T$ and $\overset{*}{z}{}^2 = (1, 0)^T$ as given in the first equilibrium.

(ii) Profit maximization: Only the consumer 1 (2) takes part of the general meeting of the firm 2 (1). Thus the firm evaluates the production plan using only the gradient Π^i of its single owner. Since $\Pi_0^i = 0$, again only the figure in (a) is relevant: Firm 1 chooses a production plan on the set Y^1, reaching for the highest possible indifference curve of consumer 2. Firm 2 chooses analogously. Thus $\overset{*}{y}{}^1 = (-1, 0, 1)^T,\ \overset{*}{y}{}^2 = (-1, 1, 0)^T$.

Analogously the second Drèze equilibrium can be found.

(c) The second Drèze equilibrium is limited Pareto-better than the first one. Thus the first Drèze equilibrium is not limited Pareto-efficient. Obviously there can be a "wrong" ownership structure.

One may rightly comment that the depicted economy is not nice in the sense that it doesn't fulfil the assumption $\mathcal{U}$ (indifference curves shouldn't intersect with the axes). One could suspect that this assumption contributes to the Pareto efficiency of Drèze equilibria, but this is not the case.

6.8 By definition in Exercise 6.8 $\overset{*}{\pi}^j$ satisfies NAC for j we have:

$$\overset{*}{\pi}^{jT} d^j = \overset{*}{\pi}^{jT} y^j + \overset{*}{\pi}^{jT} \begin{pmatrix} -q^T \\ A \end{pmatrix} \zeta^j + \overset{*}{\pi}^{jT} \begin{pmatrix} (D_0^T - p)^T \\ D_1 \end{pmatrix} \tau^j = \overset{*}{\pi}^{jT} y^j .$$

Thus the financial policy of the firms is irrelevant for them. Irrelevance for the consumer:

Consider a $\bar{x}$ being part of $\mathbb{B}^i(\overset{*}{q}, \overset{*}{p}, \omega^i, \bar{\delta}^i, A, \overset{*}{D})$ and an $\hat{x}$ being part of $\mathbb{B}^i(\overset{*}{q}, \overset{*}{p}, \omega^i, \bar{\delta}^i, A, \hat{D})$.
Thus,

$$\bar{x} = \omega^i + \begin{pmatrix} -\overset{*}{q}^T \\ A \end{pmatrix} \overset{*}{z}^i + \begin{pmatrix} (\overset{*}{D}_0^T - \overset{*}{p})^T \\ \overset{*}{D}_1 \end{pmatrix} \overset{*}{\delta}^i + \begin{pmatrix} \overset{*}{p}^T \bar{\delta} \\ 0 \end{pmatrix},$$

and

$$\hat{x} = \omega^i + \begin{pmatrix} -\overset{*}{q}^T \\ A \end{pmatrix} \hat{z}^i + \begin{pmatrix} (\hat{D}_0^T - \overset{*}{p})^T \\ \hat{D}_1 \end{pmatrix} \hat{\delta}^i + \begin{pmatrix} \overset{*}{p}^T \bar{\delta} \\ 0 \end{pmatrix},$$

where

$$\overset{*}{D} = [\overset{*}{y} + \begin{pmatrix} -\overset{*}{q}^T \\ A \end{pmatrix} \overset{*}{\zeta} - \begin{pmatrix} \overset{*}{p}^T \overset{*}{\tau} \\ 0 \end{pmatrix}][I - \overset{*}{\tau}]^{-1},$$

$$\overset{*}{D} = [\overset{*}{y} + \begin{pmatrix} -\overset{*}{q}^T \\ A \end{pmatrix} \hat{\zeta} - \begin{pmatrix} \overset{*}{p}^T \hat{\tau} \\ 0 \end{pmatrix}][I - \hat{\tau}]^{-1}.$$

Now choose $\overset{*}{\zeta}^i$ and $\hat{\zeta}$ such that:

$$[I - \overset{*}{\tau}]^{-1} \overset{*}{\zeta}^i = [I - \hat{\tau}^i]^{-1} \hat{\zeta}^i := h.$$

Then it remains to choose $\overset{*}{z}^i$ and $\hat{z}^i$ such that:

$$\begin{pmatrix} -\overset{*}{q}^T \\ A \end{pmatrix} (\overset{*}{Z}^i + \overset{*}{Z}h) - \overset{*}{p}^T (\overset{*}{\tau}h + \overset{*}{\tau}^i) \begin{pmatrix} 1 \\ 0 \\ . \\ . \\ 0 \end{pmatrix} = \begin{pmatrix} -\overset{*}{q}^T \\ A \end{pmatrix} (\hat{Z}^i + \hat{Z}h) - \overset{*}{p}^T (\hat{\tau}h + \hat{\tau}^i) \begin{pmatrix} 1 \\ 0 \\ . \\ . \\ 0 \end{pmatrix}.$$

Thus,

$$\hat{\zeta}^i = [I - \hat{y}][I - \overset{*}{\tau}^i]^{-1}\overset{*}{\zeta}^i,$$

$$\hat{Z}^i = \overset{*}{Z}^i + [\overset{*}{Z} - \hat{Z}][I - \overset{*}{\tau}]^{-1}\overset{*}{\zeta}^i,$$

are the “undo” strategies.

7 Information Asymmetries on Financial Markets

7.1

(a) Demand of consumer 1

$$\max_{\theta_1^1, \theta_2^1} \ln(x_0^1) + \frac{1}{2}\ln(x_1^1) + \frac{1}{2}\ln(x1_2^1)$$

$$x_0^1 + q_1\theta_1^1 + q_2\theta_2^1 = 1$$

$$x_1^1 = \theta_1^1,\ x_2^1$$

$$\Longrightarrow\ x_0^1 = \frac{1}{2},\ x_1^1 = \frac{1}{4q_1},\ x_2^1 = \frac{1}{4q_2}$$

Case $s = 1$ revealed to $i = 2$

$$\max_{\theta_1^2,\ (\theta_2^2=0)} \ln(x_0^2) + \ln(x_1^2)$$

$$x_0^2 + q_1\theta_1^2 = 0$$

$$x_1^2 = 1 + \theta_1^2$$

$$\Longrightarrow\ x_0^2 + q_1(x_1^2 - 1) = 0 \Leftrightarrow x_0^2 = q_1(1 - x_1^2) \tag{7.1}$$

$$\max_{x_1^2} \ln(q_1(1 - x_1^2)) + \ln(x_1^2)$$

T. Hens, M. O. Rieger, *Solutions to* Financial Economics,
Springer Texts in Business and Economics,
https://doi.org/10.1007/978-3-662-59889-4_15

$$-\frac{1}{1-x_1^2}+\frac{1}{x_1^2}=0$$

$$x_1^2=1-x_1^2 \implies x_1^2=\frac{1}{2}$$

Entering the result in Eq. (7.1) provides $x_0^2=\frac{q_1}{2}$
$\implies$ no equilibrium possible:

$$x_0^1+x_0^2=1 \tag{7.2}$$

$$\frac{1}{2}+\frac{q_1}{2}=1 \implies q_1=1 \tag{7.3}$$

$$x_1^1+x_1^2=1 \tag{7.4}$$

$$\frac{1}{4q_1}+\frac{1}{2}=1 \implies q_1=\frac{1}{2} \tag{7.5}$$

contradiction between (7.3) and (7.5). Entering the result in Eq. (7.3) provides:

$$x_2^1+x_2^2=1$$

$$\frac{1}{4q_2}+0=1 \implies q_2=\frac{1}{4}$$

Case $s=2$ revealed to $i=2$.
By symmetry:

$$x_0^2=\frac{q_2}{2}, x_2^2=\frac{1}{2}, x_1^2=0$$

$\implies$ no equilibrium possible:

$$\frac{1}{2}+\frac{q_2}{2}=1 \implies q_2=1 \tag{7.6}$$

$$\frac{1}{4q_1}+0=1 \implies q_1=\frac{1}{4} \tag{7.7}$$

$$\frac{1}{4q_2}+\frac{1}{2}=1 \implies q_2=\frac{1}{2} \tag{7.8}$$

contradiction between (7.6) and (7.8).

(b) Conjecture $\overset{*}{x}^1 = (\frac{1}{2}, \frac{1}{2}, \frac{1}{2}) = \overset{*}{x}^2$. Market clearing ✓.
Utility gradients

$$\nabla U^1(\overset{*}{x}^1) = \begin{pmatrix} 2 \\ 1 \\ 1 \end{pmatrix} = \nabla U^2(\overset{*}{x}^2)$$

Asset prices

$$\overset{*}{q}^T = \frac{1}{2}\nabla_1 U^i(\overset{*}{x}^i)^T \begin{pmatrix} 1 & 0 \\ 0 & 1 \end{pmatrix} = (\frac{1}{2}, \frac{1}{2})$$

Asset allocation

$$\overset{*}{\theta}^1 = (\frac{1}{2}, \frac{1}{2}) = -\overset{*}{\theta}^2$$

Budget restrictions

$$(i = 1) \quad \frac{1}{2} + \frac{1}{2} \quad t = 0$$

$$\frac{1}{2} = A\overset{*}{\theta}^1 = \frac{1}{2} \quad s = 1, 2$$

$$(i = 2) \quad -\frac{1}{2} + \frac{1}{2} = 0 \quad t = 0$$

$$\frac{1}{2} = 1 - \frac{1}{2} \quad s = 1, 2$$

First-order-condition satisfied by definitions of asset prices.

(c) Consumer 1 understands the following:

$q_1 = q_2 \implies$ consumer 2 is not informed (equilibrium as in (b))
$q_1 > q_2 \implies$ consumer 2 knows $s = 1$ occurs demand of consumer 1 in that case:

$$\max_{\theta_1^1,\ (\theta_2^1 = 0)} \ln(x_0^1) + \ln(x_1^1)$$

$$x_0^1 + q_1\theta_1^1 = 1, x_1^1 = \theta_1^1, x_2^1 = 0$$

$$\max_{x_1^1} \ln(1 - q_1 x_1^1) + \ln(x_1^1)$$

$$\frac{-q_1}{1 - q_1 x_1^1} + \frac{1}{x_1^1}$$

$$\frac{1}{x_1^1} = \frac{q_1}{1 - q_1 x_1^1}$$
$$1 - q_1^1 x_1^1 = q_1 x_1^1$$
$$1 = 2q_1 x_1^1$$
$$x_1^1 = \frac{1}{2q_1}, x_0^1 = \frac{1}{2}$$

from 7.1(a) follows $x_0^2 = \frac{q_1}{2},\ x_1^2 = \frac{1}{2}$,

$$\frac{1}{2} + \frac{q_1}{2} = 1 \implies q_1 = 1$$
$$\frac{1}{2q_1} + \frac{1}{2} = 1 \implies q_1 = 1$$

equilibrium is $q_1 = 1, q_2 < 1$ allocation as before $\frac{1}{2}$ each.
$q_1 < q_2 \implies$ consumer 2 knows $s = 2$ occurs by symmetry:

$$q_2 = 1,\ q_1 < 1,$$

allocation as before: $\frac{1}{2}$.

7.2 (Extracted from [HST97], solution 13.B.4)

We think of the true valuation of the good, y, as a state of nature s $\in$ S where S is the set of all states of nature, and both agents 1 (say the seller) and 2 (say the buyer) have a common prior that is common knowledge. Let $H_i(s)$ be the set of states that agent i believes is possible given the true states s (this will depend on the signal observed). We can define the following two events:

$$T_1 \equiv s \in S \mid E[y \mid H_1(s) \cap T_2] \leq p T_2 \equiv s \in S \mid E[y \mid H_2(s) \cap T_1] \geq p$$

That is, T_i is the event that agent i will say "trade" given that he knows that event $H_i(s)$ has occurred, and agent j will say "trade". Assume that there exists an equilibrium where both agents say "trade". Then, each agent i *knows* that event T_i occurred, and since this is an equilibrium then each agent i *belives* with probability 1 that event $T = T_1 \cap T_2$ occurred. The seller (1) then prefers to trade if and only if $E[y \mid T] \leq p$, and the buyer (2) prefers to trade if and only if $E[y \mid T] \geq p$. Assuming a generic distribution of values both can hold with probability zero, therefore the set T must occur with probability zero.

7.3 (Extracted from [Tir10], solution 3.12)

(a) This condition says that the risk-free rate is normalized at 0. In other words, investors are willing to lend 1 unit at date 0 against a safe return of 1 unit at date 1.

(b) With a competitive capital market, the financing condition becomes

$$p_H q_S R_1 \geq I - A.$$

With a risk-neutral entrepreneur, the incentive compatibility constraint is unchanged:

$$(\Delta p) R_b \geq B.$$

Thus, enough pledgeable income can be harnessed provided that

$$p_h \left[R - \frac{B}{\Delta p} \right] \geq \frac{I - A}{q_S}. \tag{7.9}$$

We conclude that obtaining financing is easier for a countercyclical firm than for a procyclical one, ceteris paribus.

(c) The entrepreneur maximizes her utility subject to the investors' being willing to lend

$$\max_{\{R_B^F, R_b^S\}} \{p_H R_b^S + (1 - p_H) R_b^F\} \tag{7.10}$$

s.t.

$$q_S p_H (R - R_b^S) + q_F (1 - p_H)(-R_b^F) \geq I - A, \tag{7.11}$$

$$(\Delta p)(R_b^S - R_b^F) \geq B, \tag{7.12}$$

$$R_b^F \geq 0. \tag{7.13}$$

Letting μ_1, μ_2, and μ_3 denote the shadow prices of the constraints, the first-order conditions are

$$p_H [1 - \mu_1 q_S] + \mu_2 (\Delta p) = 0 \tag{7.14}$$

and

$$(1 - p_H)[1 - \mu_1 q_F] - \mu_2 (\Delta p) + \mu_3 = 0. \tag{7.15}$$

- First, note that for $q_S \neq q_F$ at least one of constraints (7.12) and (7.13) must be binding: if $\mu_2 = \mu_3 = 0$, (7.14) and (7.15) cannot be simultaneously satisfied.
- Conversely, (7.12) and (7.13) cannot be simultaneously binding, except when condition (7.9) is satisfied with exact equality.
- Suppose that constraint (7.12) is not binding ($\mu_2 = 0$), which, from what has gone before, implies that $R_b^F = 0$. Then $\mu_1 = 1/q_s$, and (7.15) can be

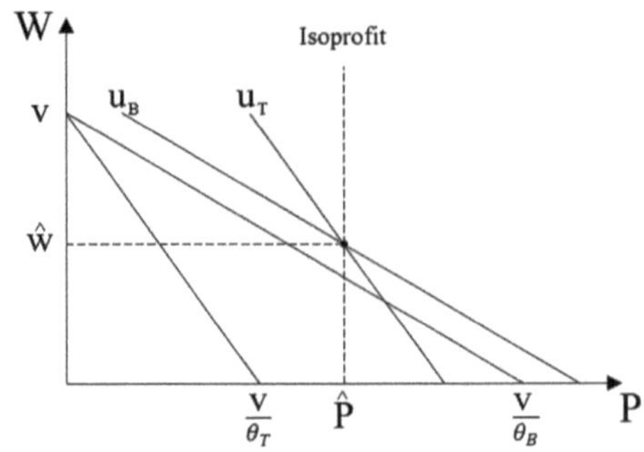

Fig. 1 Waiting time is on y-axis and Price on x-axis

satisfied only if

$$q_F > q_S.$$

- In contrast, suppose that constraint (7.13) is not binding ($\mu_3 = 0$). Constraints (7.14) and (7.15) taken together imply that

$$q_s > q_F.$$

To sum up, the maximum punishment result ($R_b^F = 0$) carries over to procyclical firms, because the incentive effect compounds with the "marginal rates of substitution" effect (the investors value income in the case of failure relatively more compared with the entrepreneur). But it does not in general hold for countercyclical firms. Then the investors care more about the payoff in the case of success, and the entrepreneur should keep marginal incentives equal to $B/\Delta p$ and select $R_B^F > 0$ (since the firm's income is equal to 0 in the case of failure, this requires the firm to hoard some claim at date 0 so as to be able to pay the entrepreneur even in the case of failure).

Entrepreneurial risk aversion changes the incentive constraint (7.12) and the objective function (7.10). It may be the case that $R_b^F > 0$ even for a procyclical firm.

7.4 (Extracted from [HST97], solution 14.C.8)

(a) The indifference curves of the two types and the firm's isoprofit curve are depicted in Fig. 1. The formal problem that Air Shangri-La would want to solve is:

$$\max_{\{P_T, W_T, P_B, W_B\}} \quad \lambda P_T + (1-\lambda) P_B$$

$$\text{s.t.} \quad \theta_T P_T + W_T \leq v \tag{7.16}$$

$$\theta_B P_B + W_B \leq v \tag{7.17}$$

$$\theta_T P_T + W_T \leq \theta_T P_B + W_B \tag{7.18}$$

$$\theta_B P_B + W_B \leq \theta_B P_T + W_T \tag{7.19}$$

$$P_B, W_B, P_T, W_t \geq 0 \tag{7.20}$$

(b) In the program above, constraint (7.16) and (7.19), together with $\theta_B < \theta_T$, imply that constraint (7.17) is redundant, so it is never binding. Constraint (7.16) must therefore bind for if it would not, we can reduce P_T and P_B by $\epsilon > 0$, and all the remaining constraints will still be satisfied. This implies that tourists will be indifferent between buying and not buying a ticket.

(c) Assume that $\{(P_T, W_T), (P_B, W_B)\}$ is an optimal, incentive compatible contract, and assume in negation that $W_B > 0$. Now reduce W_B by $\epsilon > 0$, and increase P_B by $\frac{\epsilon}{P_B}$ so that the B type's utility does not change, and the firm earns higher profits from the B type. We need to check that the T type will not choose this new compensation package. Indeed,

$$\theta_T P_T + W_T \leq \theta_T P_B + W_B =$$

$$\theta_T (P_B + \frac{\epsilon}{\theta_T}) + (W_B - \epsilon) < \theta_T (P_B + \frac{\epsilon}{\theta_B}) + (W_B - \epsilon),$$

contradicting that $\{(P_T, W_T), (P_B, W_B)\}$ is an optimal, incentive compatible contract. Therefore, we must have $W_B = 0$. If, in an optimal contract, the business travelers were not indifferent between $(P_T.W_T)$ and (P_B, W_B), we could slightly raise P_B and all the constraints would remain satisfied (recall that (7.17) is redundant), and the firm would earn higher profits from the business types. Therefore, in an optimal contract we must have the business types indifferent between (P_T, W_T) and (P_B, W_B).

(d) The trade off that the firm faces is: By lowering P_T and increasing W_T so as to keep the tourists indifferent between buying a ticket or not, the firm can increase P_B (recall that $W_B = 0$). From parts (b) and (c) above, we can conclude that if the firm raises W_T by ϵ and lowers P_T by $\frac{\epsilon}{\theta_T}$ so that the tourists remain indifferent between buying or not, then to keep the business types indifferent between their package and the new tourist package, it can increase P_B by $\frac{\epsilon(\theta_T - \theta_B)}{\theta_T \theta_B}$. Since this trade off is linear, it is true no matter where we are in the (P, W) space, and therefore it will be profitable if and only if:

$$\lambda \frac{\epsilon}{\theta_T} < (1 - \lambda) \frac{\epsilon(\theta_T - \theta_B)}{\theta_T \theta_B},$$

or,

$$\frac{\lambda}{1 - \lambda} < \frac{\theta_T - \theta_B}{\theta_B}.$$

Note that this is independent of the cost c, because this is a revenue trade off (the costs are the same for both types). Therefore, the price discrimination scheme can take on two forms:

(a) If $\frac{\lambda}{1-\lambda} > \frac{\theta_T - \theta_B}{\theta_B}$, then only the high types will be served and the scheme will be (this assumes that $c < \frac{v}{\theta_B}$:

$$\{(P_T, W_T) = (P_B, W_B)\} = \{(0, v), (\frac{v}{\theta_B, 0})\}$$

(b) If $\frac{\lambda}{1-\lambda} > \frac{\theta_T - \theta_B}{\theta_B}$, then both types will be served and the scheme will be a pooling scheme with (this assumes that $c < \frac{v}{\theta_T}$):

$$(P_T, W_T) = (P_B, W_B) = (\frac{v}{\theta_T}, 0).$$

Since the direction of the inequality in the condition above determines the type of scheme, it is easy to see how changes in λ, θ_T, and θ_B will affect the scheme: If the proportion of B types is large enough (λ small enough) the firm will choose to serve only the B types. If the B types suffer less from prices (θ_B is smaller) then the firm is more likely to serve only them. If the T types suffer less from prices (θ_T is smaller) then the firm is more likely to serve them as well as the B types. Changes in the cost c are discussed below.

(e) As long as $c < \frac{v}{\theta_B}$, and we are in case 1 as described in part (d) above, the firm will decide to serve only the business types. If, however, we are in case 2 above, and $\frac{v}{\theta_T} < c < \frac{v}{\theta_B}$, then the scheme described in (d) above cannot be optimal because the firm is losing money. In such a case, the firm will choose the scheme described in case 1, and serve only the business types. If $c > \frac{v}{\theta_B}$ the firm will choose not to operate at all.

7.5 (Extracted from [HST97], solution 14.C.9)

(a) The monopolist will offer the individual a policy that fully insures him (optimal risk sharing) and keeps the individual at the same level of expected utility. That is, if we define $\bar{o} \equiv \theta u(W - L) + (1 - \theta)u(W)$, then the optimal insurance policy has $c_1 = c_2 = u^{-1}(\bar{u})$.

(b) The monopolist will offer an optimal screening contract of the form $(c_1^L, c_2^L), (c_1^H, c_2^H)$ that solves:

$$\begin{aligned}
\text{Max}\quad & \lambda[(1-\theta_H)c_1^H + \theta_H c_2^H] + (1-\lambda)[(1-\theta_L)c_1^L + \theta_L c_2^L] \\
\text{s.t.}\quad & (1-\theta_H)u(c_1^H) + \theta_H u(c_2^H) \geq \bar{u} \\
& (1-\theta_L)u(c_1^L) + \theta_L u(c_2^L) \geq \bar{u} \\
& (1-\theta_H)u(c_1^H) + \theta_H u(c_2^H) \geq (1-\theta_H)u(c_1^L) + \theta_H u(c_2^L) \\
& (1-\theta_L)u(c_1^L) + \theta_L u(c_2^L) \geq (1-\theta_L)u(c_1^H) + \theta_L u(c_2^H)
\end{aligned}$$

This is again a standard monopolistic screening problem and the standard analysis applies. Solving this program will have the H type fully insured, and the L type not fully insured. A graphical analysis is given using figure below. Both types start at the point A, with utility levels $\bar{u}_L$ and $\bar{u}_H$ respectively. If the monopolist would try to insure both types ba offering the point B and C to H and L respectively, the (unobservable) H type would choose policy C instead of B, that is, the points B and C cannot be part of an incentive compatible contract. If the monopolist offers points A and B, then the H types would choose B (since they are indifferent), they would be fully insured, and the risk neutral monopolist would make profits from their choice. The L types, however, would prefer the point A to B and no profits would be made from them. If the proportion of High types is large enough, then the monopolist will find it profitable to slightly insure the L type at the cost of raising the utility of the H type. This means that the optimal contract will look like the two points D and E for the H and L types respectively. The mathematical analysis is straightforward, and results in a situation common to monopolistic screening (or hidden information): The H type will be at the first best insurance level and his participation constraint is not binding. The L type will be under insured (second best distortion so that screening is possible and profitable), his participation constraint will bind and his incentive compatibility constraint will not. (Note, that the pair of points A,B may be optimal if the proportion of l types is small. this is parallel to the hidden information case where the H type is at the first-best observable point with no surplus, and the L type has $e_L = 0$).

(c) The difference is who gets the surplus. In comparison to competitive markets, where the insurer was left with zero profits and the individuals had utility levels above their reservation utility. Here, the monopolist makes positive profits and at least the L type has no surplus.

7.6 (Extracted from [Tir10], solution 3.21)

(a) Letting R_b denote the entrepreneur's stake in success (and 0 in failure), the incentive compatibility constraint is

$$(\Delta p)R_b \geq B.$$

Financing is feasible if and only if

$$p_H(R - \frac{B}{\Delta p}) \geq I - A.$$

The entrepreneur's date-1 *gross* utility is

$$[p_H R - I] + [a - \bar{A}] if A \geq \bar{A}.$$

- If $A_0 \geq \bar{A}$, the entrepreneur's date-0 expected gross utility is

$$U_b^h = [p_H R - I] + A_0$$

if she hedges.

- By contrast, and letting $F(\varepsilon)$ denote the cumulative distribution of ε, her expected utility becomes

$$U_b^g = [1 - F(\bar{A} - A_0)][[p_H R - I] + m^+(\bar{A})] + F(\bar{A} - A_0)m^-(\bar{A}) < U_b^h,$$

where

$$m^+(\bar{A}) \equiv E[A \mid A \geq \bar{A}],$$

$$m^-(\bar{A}) \equiv E[A \mid A < \bar{A}],$$

$$[1 - F(\bar{A} - A_0)]m^+(\bar{A}) + F(\bar{A} - A_0)m^-(\bar{A}) = A_0.$$

- If $A_0 < \bar{A}$, then

$$U_b^h = A_0 < U_b^g.$$

(b) *Ex post* the entrepreneur chooses p so as to solve

$$\max_{\{p\}}\{pR_b - \frac{1}{2}p^2\},$$

and so

$$p = R_b.$$

The pledgeable income is

$$P = R_b(R - R_b)$$

and the NPV, i.e., the entrepreneur's expected *net* utility, in the case of financing is

$$U_b = R_b R - I.$$

Without loss of generality, assume that $R_b \geq \frac{1}{2}R$ (if $R_b < \frac{1}{2}$, $\hat{R}_b = R - R_b$ yields the same P and a higher U_b).

Assume that $I - A_0 < \frac{1}{4}R^2$. This condition means that the entrepreneur can receive funding if she hedges (the highest pledgeable income is reached for

$R_b = \frac{1}{2}R$). She also receives funding even in the absence of hedging provided that the support of ε is small enough (the lower bound is smaller than $\frac{1}{4}R^2 - (I - A_0)$ in absolute value). Let

$$V(A) \equiv R_b(A)R - I,$$

where $R_b(A)$ is the largest root of

$$R_b(R - R_b) = I - A.$$

One has

$$\frac{dV}{dA} = R\frac{dR_b}{dA} = \frac{R}{2R_b(A) - R} > 0$$

and

$$\frac{d^2V}{dA^2} = -\frac{2R}{(2R_b(A) - R)^2}\frac{dR_b}{dA} < 0.$$

Hence, V is concave and so

$$V(A_0) > E[V(A_0 + \varepsilon)].$$

The entrepreneur is better off hedging.

(c) The investment is given by the investors' breakeven condition:

$$p_H[RI - \frac{B(I)}{\Delta}] = I - A.$$

This yields investment $I(A)$, with $I' > 0$ and $I'' < 0$ if $B'' > 0$, $I'' > 0$ if $B'' < 0$. The *ex ante* utility is

$$U_b^h = (p_H R - 1)E[I(A_0 + \varepsilon)]$$

in the absence of hedging. And so $U_b^h < U_b^g$ if $B'' < 0$.

(d) When the profit is unobservable by investors, there is no pledgeable income and so

$$I = A.$$

And so

$$U_b^h = R(A_0) \text{ and } U_b^g = E[R(A_0 + \varepsilon)] < R(A_0)$$

since R is concave.

(e) Quite generally, in the absense of hedging the realization of ε generates a distribution $G(I)$ over investment levels $I = I(\varepsilon)$ and over cash used in the project $A(\varepsilon) \leq A_0 + \varepsilon$ such that

$$P(I(\varepsilon)) \geq I(\varepsilon) - A(\varepsilon),$$

where P is the pledgeable income. And so

$$E[P(I)] \geq E[I] - A_0.$$

Drawing I from distribution $G(.)$ regardless of the realization of ε and keeping $A_0 - E[A(\varepsilon)]$ makes the entrepreneur as well off.
In general, the entrepreneur can do strictly better by insulating her investment from the realization of ε.

Consider, for example, the case $A_0 < \bar{A}$ in subquestion (a). Then we know that gambling is optimal. The probability that the project is financed is

$$1 - F(\bar{A} - A_0) \text{ and } [1 - F(\bar{A} - A_0)]\bar{A} < A_0.$$

This last inequality states that there is almost surely "unused cash": either $A_0 + \varepsilon < \bar{A}$ and then there is no investment, or $A_0 + \varepsilon > \bar{A}$ and then there is "excess cash" $[A_0 + \varepsilon - \bar{A}]$. Consider therefore the date-0 contract in which the date-1 income $r = A_0 + \varepsilon$ is pledged to investors. The probability of funding is the X, which allows investors to break even:

$$A_0 = X[I - p_H(R - \frac{B}{\Delta p})] = X\bar{A}.$$

Clearly,

$$X > 1 - F(\bar{A} - A_0),$$

and so the entrepreneur's date-0 expected gross utility has increased from

$$[1 - F(\bar{A} - A_0)](p_H R - I) + A_0$$

to

$$X(p_H R - I) + A_0.$$

Of course, this is not quite a fair comparison, since we have allowed random funding under hedging and not under gambling. But, because there is excess cash in states of nature which $A > \bar{A}$, the project could be funded with probability $x(A) = A/\bar{A}$. The total probability of funding under gambling would then be

$$\int_0^{\bar{A}-A_0} \frac{AdF(A-A_0)}{\bar{A}} + [1 - F(\bar{A} - A_0)]$$

$$< \frac{\int_0^{\bar{A}-A_0} AdF(A-A_0) + \int_{\bar{A}-A_0}^{\infty} AdF(A-A_0)}{\bar{A}} = \frac{A_0}{\bar{A}}.$$

7.8 (Extracted from [HST97], solution 13.D.2)

(a) Once M and R are given, the wealth levels of an insured individual in the two states, denoted by (w_1, w_2), are given by:

$$(w_1, w_2) = (W - M, W - L - M + R).$$

Therefore, we can think of the insurance contract as specifying the wealth levels (w_1, w_2) in the two states. The premium M and the repayment R can be obtained by the following equations: $M = W - w_1$ and $R = w_2 + L - w_1$.

(b) This game is analogous to the screening game studied in this chapter. Therefore there exists no pooling equilibrium, and the existence of a separating equilibrim is not always assured. If there exists a separating equilibrium, then the high risk types are completely insured, i.e. $w_1^H = w_2^H$.
The low risk types will not be completely insured, in fact $w_1^L > w_2^L$.

7.9 (Extracted from [HST97], solution 6.1)

(a) If the borrower's privat benefit B were common knowledge, then, if financed, the borrower would receive R_b in the case of success, with

$$R_b \geq \frac{B}{\Delta p},$$

so as to include her to behave. The project would be funded if and only if the pledgeable income exceeded the investment cost:

$$p_H(R - \frac{B}{\Delta P}) \geq I.$$

Suppose that the borrower offers a contract specifying that she will receive R_b in the case of success and 0 in the case of failure (offering to receive more than 0 in the case of failure would evidently raise suspicion, and can indeed be shown not to improve the borrower's welfare). There are three possible cases:

(i) $R_b \geq \frac{B_H}{\Delta p}$ induces the borrower to work regardless of her type, and thus creates an information insensitive security for the lenders, who obtain

$$p_H(R - R_b) - I \leq p_H(R - \frac{B_H}{\Delta p}) - I < 0.$$

So, such high rewards for the borrower cannot attract financing.

(ii) $R_b < \frac{B_L}{\Delta p}$ induces the borrower to shirk regardless of her type. The lenders' claim is again informative insensitive, and fails to attract financing.

(iii) $\frac{B_L}{\Delta p} \leq R_b < \frac{B_H}{\Delta p}$: suppose that, in equilibrium, the good borrower offers a contract with a reward in this range, and that this attracts financing. A bad borrower must then then "pool" and offer the same contract: if she were to offer a different contract, her type would be revealed to the capital market and her project would not be funded. Furthermore, she receives utility from the project being funded at least equal to that of a good borrower (she receives the same payoff conditional on working and a higher payoff conditional on shirking). So, she is better off pooling with the good borrower than not being funded.

We conclude that equilibrium is necessarily a pooling equilibrium. It either involves no funding at all or funding of both types. From the study of cases (i) and (ii), we also know that, in the case of funding, the good type behaves and the bad one misbehaves.

(b) A necessary condition for funding is thus that

$$[\alpha p_H + (1 - \alpha)p_L](R - R_b) \geq I.$$

Since $R_b \geq \frac{B_L}{\Delta p}$, there cannot be any lending if

$$\alpha < \alpha^*,$$

where

$$[\alpha^* p_H + (1 - \alpha^*)p_L](R - \frac{B_L}{\Delta p}) = I.$$

Thus, if the proportion of good borrowers is smaller than $\alpha^* \in (0, 1)$, there is no lending at all. Bad borrowers drive out good ones and the loan market breaks down.

Suppose, next, that the proportion of good borrowers is high: $\alpha > \alpha^*$. The borrower may now be able to receive financing. Suppose that the borrower, regardless of her type, offers to receive R_b^* in the case of success and 0 in the case of failure, where

$$[\alpha p_H + (1 - \alpha)p_L](R - R_b^*) = I.$$

Because $\alpha > \alpha^*$, $R_b^* > \frac{B_L}{\Delta p}$ and so the good borrower behaves. The investors' breakeven condition is therefore satisfied. It is an equilibrium for both types to offer contract $\{R_b^*, 0\}$ and for the capital market to fund the project.

(c) The pooling equilibrium (which exists whenever $\alpha \geq \alpha^*$) exhibits no market breakdown. Indeed, there is more lending under adverse selection than under symmetric information.

It involves an externality between the two types of borrower. The good type obtains reward

$$R_b^* = \frac{R - I}{[\alpha p_H + (1 - \alpha) p_L]}$$

in the case of success below that, $\frac{R-I}{p_H}$, that she would obtain under symmetric information.

The project's NPV conditional on being funded falls from $p_H R - I$ to $[\alpha p_H + (1 - \alpha) p_L] R - I$ due to asymmetric information. The quality of lending is thus affected by adverse selection.

7.10 (Extracted from [HST97], solution 6.6)

(a) Condition (1) means that the pledgeable income of a good (bad) borrower exceeds (is lower than) the investors' investment $I - A$. The pledgeable income is equal to the expected income, $p_H R$, minus the entrepreneur's incompressible share, $\frac{p_H b}{\Delta p}$ (or $\frac{p_H B}{\Delta p}$).

To see that no lending occurs in equilibrium, note that the bad type (type B) always derives a (weakly) higher surplus from being financed than a good type (type B). Hence, contracts that provide financing to a good type will also provide financing to a bad one (pooling behaviour).

Condition (1) implies that one cannot offer a breakeven contract must induce misbehavior by the bad type. But condition (2) in turn implies that pooling contracts with stakes for the borrower in the interval $(\frac{b}{\Delta p}, \frac{B}{\Delta p})$ generate a loss for the investors.

(b) In a separating equilibrium the good type chooses x and then offers R_b, and the bad type, which is recognized, chooses $x = 0$ and, from condition (1), receives no funding. Were the bad type to mimic the good type, she would get funding with probability $1 - x$; for, either the signal reveals the type and then she gets no funding, or the signal reveals nothing and the investors still believe they face a good type (we here use the fact that the equilibrium is separating).

Letting R_b^G denote the good type's "full information" (with net capital $A - rx$) contract (given by $p_H(R - R_b^G) = I - A + rx$), it must be the case that the bad type does not want to mimic the good type and prefers to keep her capital A instead. That is,

$$A \geq (1 - x)[p_L R_b^G + B] + x(A - rx)$$

or

$$A \geq x(A - rx) + (1 - x)[p_L(R - \frac{I - A + rx}{p_H}) + B],$$

which yields the condition in the question. This condition is satisfied with equality at the separating equilibrium.

8 Time-Continuous Model

8.1

$$d(W^2) = 2WdW + \frac{1}{2}^2 (dW)^2 = 2WdW + dt.$$

8.2 Taking $T \to 0$ we obtain:

$$df(t) = \frac{1}{2}\sigma^2 f(t)dt$$

$$\frac{df(t)}{f(t)} = \frac{1}{2}\sigma^2 dt$$

$$d\log f(t) = \frac{1}{2}\sigma^2 dt$$

Integrating from 0 to T gives:

$$\log f(t) - \log f(0) = \frac{1}{2}\sigma^2 T$$

$$\log \frac{f(T)}{f(0)} = \frac{1}{2}\sigma^2 T$$

Taking the exponential and rearranging:

$$f(T) = f(0)e^{\frac{1}{2}\sigma^2}$$

$$E[e^{\mu+\sigma W(T)}] = e^{\mu+\frac{1}{2}\sigma^2 T}$$

T. Hens, M. O. Rieger, *Solutions to* Financial Economics,
Springer Texts in Business and Economics,
https://doi.org/10.1007/978-3-662-59889-4_16

In particular,

$$E[e^{W(T)}] = e^{T/2} = e^{\text{VAR}(W(T))/2}$$

8.3

(a) By Ito's Lemma, we have:

$$\begin{aligned} d\log S &= \frac{1}{S}dS - \frac{1}{2}\frac{1}{S^2}(dS)^2 \\ &= \frac{dS}{S} - \frac{1}{2}\sigma^2 dt \\ &= (\mu - \frac{1}{2}\sigma^2)dt + \sigma dW \\ S(T) &= S(0)\cdot e^{(\mu-\frac{1}{2}\sigma^2)T+\sigma W} \end{aligned}$$

Thus:

$$\log S(t) - \log S(0) = (\mu - \frac{1}{2}\sigma^2)T + \sigma W,$$

the left hand side equals $\log \frac{S(T)}{S(0)}$ and taking the exponential thus gives

$$S(T) = S(0)\cdot e^{\mu-\frac{1}{2}\sigma^2 T+\sigma W}$$

Hence

$$\begin{aligned} E[S(T)] &= S(0)\cdot e^{\mu-\frac{1}{2}\sigma^2 T+\frac{\sigma^2 T}{2}} \\ &= S(0)\cdot e^{\mu T}. \end{aligned}$$

(b) We have seen in (a) that

$$\log S(t) = \log S(0) + (\mu - \frac{1}{2}\sigma^2)T + \sigma W(T)$$

Taking the expected value gives

$$E[\log]S(t) = \log S(0) + (\mu - \frac{1}{2}\sigma^2)T,$$

and taking the variance gives

$$\text{VAR}[\log]S(t) = \sigma^2 T,$$

which leads finally to

$$\sqrt{\mathrm{VAR} \log S(t)} = \sigma \sqrt{T}.$$

(c) We calculate

$$\begin{aligned} E[S^2(T)] &= E[(S(0) \cdot e^{(\mu - \frac{1}{2}\mu^2)T + \sigma w(T)})^2] \\ &= S^2(0) \cdot e^{(\mu - \frac{1}{2}\sigma^2)2T + \frac{\sigma^2 4T}{2}} \\ &= S^2(0) \cdot e^{2\mu T + \sigma^2 T} \end{aligned}$$

8.4 Let $h_t := \sqrt{V_t}$ then $h_t^2 = V_t$. By Ito's Lemma

$$\begin{aligned} dh_t^2 &= 2\, h\, dh + (dh)^2 \\ &= 2\sqrt{V}(-\beta\sqrt{V_t}\, dt + \delta\, dW) + \delta^2 dt \\ &= (\delta^2 - 2\beta\, V_t)dt + 2\delta\, \sqrt{V_t}\, dW. \end{aligned}$$

8.5

(a) Using Ito's Lemma, we have

$$\begin{aligned} dS^5 &= 5S^4 dS + \frac{20}{2} S^3 (dS)^2 \\ &= 5S^4(\mu S dt + \sigma S dW) + 10 S^3 \sigma^2 S^2 dt \\ &= 5S^5 \mu dt + 10 S^5 \sigma^2 dt + 5\sigma S^5 dW \end{aligned}$$

Therefore

$$\begin{aligned} \frac{dS^5}{S^5} &= 5\mu dt + 10\sigma^2 dt + 5\sigma dW \\ &= (5\mu + 10\sigma^2)dt + 5\sigma dW \end{aligned}$$

follows, which means that S^5 follows a geometric Brownian motion (with drift $5\mu + 10\sigma^2$ and volatility 5σ).

(b) A computation, similar to 8.3(c) gives:

$$\begin{aligned} E[S^5(T)] &= E[e^{(\mu - \frac{1}{2}\sigma^2)5T + \sigma 5W}] \\ &= e^{(\mu - \frac{1}{2}\sigma^2)5T + \frac{25\sigma^2}{2} T}. \end{aligned}$$

(c) We can directly apply the last formula to get

$$\begin{aligned} price_t(S^5(T)) &= \frac{1}{e^{r(T-t)}} E_t^Q[S^5(T)] \\ &= \frac{1}{e^{r(T-t)}} e^{(T-\frac{1}{2}\sigma^2)5(T-t)+\frac{25\sigma^2}{2}(T-t)}. \end{aligned}$$

8.6 This follows directly from applying Ito's Lemma:

$$\begin{aligned} dE(S(t)) &= [-\frac{1}{2}S^2(t)\sigma^2 dt]E(S(t)) + E(S(t))dS(t) + \frac{1}{2}E(S(t))S^2(t)\sigma^2 dt \\ &= E(S(t))dS(t). \end{aligned}$$

8.7

(a) The following iterative argument proves the statement:

$$\begin{aligned} M(t) &= \mathbb{E}_t[M(t+1)] \\ &= \mathbb{E}_t[\mathbb{E}_{t+1}[M(t+2)]] \\ &= \mathbb{E}_t[M(t+2)] \\ &\dots \\ &= \mathbb{E}_t[\mathbb{E}_{T-1}[M(T)]] = \mathbb{E}_t[M(T)] = 0 \end{aligned}$$

for $t = 0, \dots, T$.

(b) We compute

$$\begin{aligned} M(t) = (\sum_{s=1}^{t} R[Z_s]) - \alpha t &= \mathbb{E}_t[M(t+1)] \\ &= \mathbb{E}_t[(\sum_{s=1}^{t+1} R[Z_s]) - \alpha(t+1)] \\ &= (\sum_{s=1}^{t} R[Z_s]) + \mathbb{E}_t[R[Z_{t+1}]] - \alpha(t+1) \end{aligned}$$

The expected value of this is zero if

$$\alpha = \mathbb{E}_t[R[Z_{t+1}]] = p(1+u) + (1-p)(1+d).$$

8.8 If $\beta = 0$, the stochastic changes of the volatility σ are switched off, the second Brownian motion B_2 becomes irrelevant. The model therefore becomes very similar to the classical Black-Scholes model, but initially the volatility σ still changes: it converges exponentially to ω. To prove this, we see that the second stochastic differential equation simplifies to the ordinary differential equation

$$\sigma'(t) = \theta(\omega - \sigma(t)).$$

This equation can be easily solved as follows: define the auxiliary function $h(t) := -\omega + \sigma(t)$. We have $h'(t) = \sigma'(t)$, thus $h'(t) = -\theta h(t)$. The solution to this simple differential equation is $h(t) = h(0)e^{-\theta t}$. Converting back we obtain $\sigma(t) = \omega + (\sigma(0) - \omega)e^{-\theta t}$.

If β is on the other hand a *positive* constant, the model has still stochastic volatility, but the changes in the volatility are independent of the current volatility level. This would be rather unrealistic as well, since data clearly shows that these changes are faster when volatility is higher. The relationship between both has been studied in Sect. 8.8.4 and is illustrated in the graphic on the book cover.

References

[HST97] Chiaki Hara, Ilya Segal, and Steve Tadelis, *Solutions manual for" microeconomic theory, mas-colell, whinston, and green"*, Oxford University Press, 1997.

[MCWG+95] Andreu Mas-Colell, Michael Dennis Whinston, Jerry R Green, et al., *Microeconomic theory*, vol. 1, Oxford university press New York, 1995.

[Tir10] Jean Tirole, *The theory of corporate finance*, Princeton University Press, 2010.

T. Hens, M. O. Rieger, *Solutions to* Financial Economics,
Springer Texts in Business and Economics,
https://doi.org/10.1007/978-3-662-59889-4

MIX
Papier aus verantwortungsvollen Quellen
Paper from responsible sources
FSC® C105338

If you have any concerns about our products,
you can contact us on
ProductSafety@springernature.com

In case Publisher is established outside the EU,
the EU authorized representative is:
Springer Nature Customer Service Center GmbH
Europaplatz 3, 69115 Heidelberg, Germany

Printed by Libri Plureos GmbH
in Hamburg, Germany